Observing Variable Stars, Novae, and Supernovae

Variable stars can be fascinating objects to study. This complete practical guide and resource package instructs amateur astronomers in observing and monitoring variable stars and other objects of variable brightness. Descriptions of the objects are accompanied by explanations of the background astrophysics, providing readers with a real insight into what they are observing at the telescope. The main instrumental requirements for observing and estimating the brightness of objects by visual means and by CCD photometry are detailed, and there is advice on the selection of equipment. The book contains a CD-ROM packed with resources, including hundreds of light-curves and over 600 printable finder charts. Containing extensive practical advice, this comprehensive guide is an invaluable resource for amateur astronomers of all levels, from complete beginners to more advanced observers.

GERALD NORTH graduated in physics and astronomy. A former teacher, college lecturer, and Guest Observer at the Royal Greenwich Observatory he is now a freelance astronomer and author based in Norfolk, UK. He has been a member of the British Astronomical Association since 1977, and has served in many posts in the Lunar Section, in addition to contributing observations to various other sections. He has written numerous books, including the acclaimed *Advanced Amateur Astronomy*, and *Observing the Moon*, both published by Cambridge University Press.

Observing Variable Stars, Novae, and Supernovae

GERALD NORTH

(with accompanying CD-ROM by Nick James)

PUBLISHED BY THE PRESS SYNDICATE OF THE UNIVERSITY OF CAMBRIDGE
The Pitt Building, Trumpington Street, Cambridge, United Kingdom

CAMBRIDGE UNIVERSITY PRESS
The Edinburgh Building, Cambridge, CB2 2RU, UK
40 West 20th Street, New York, NY 10011–4211, USA
477 Williamstown Road, Port Melbourne, VIC 3207, Australia
Ruiz de Alarcón 13, 28014 Madrid, Spain
Dock House, The Waterfront, Cape Town 8001, South Africa

http://www.cambridge.org

© Gerald North 2004; CD-ROM © Nick James 2004

This book is in copyright. Subject to statutory exception
and to the provisions of relevant collective licensing agreements,
no reproduction of any part may take place without
the written permission of Cambridge University Press.

First published 2004

Printed in the United Kingdom at the University Press, Cambridge

Typefaces Palatino 10/13 pt. and Meta Plus   *System* LATEX 2$_\varepsilon$ [TB]

A catalogue record for this book is available from the British Library

Library of Congress Cataloguing in Publication data
North, Gerald.
Observing variable stars, novae, and supernovae / Gerald North (with accompanying
CD-ROM by Nick James).
   p. cm.
Includes bibliographical references and index.
ISBN 0 521 82047 2 (hardback)
1. Variable stars – Observers' manuals.   2. Stars, New – Observers' manuals.
3. Supernovae – Observers' manuals.   I. James, Nick, 1962–   II. Title.
QB835.N65   2004
523.8$'$44 – dc22   2004045760

ISBN 0 521 82047 2 hardback

The publisher has used its best endeavours to ensure that the URLs for external websites referred to in this book are correct and active at the time of going to press. However, the publisher has no responsibility for the websites and can make no guarantee that a site will remain live or that the content is or will remain appropriate.

Contents

Preface		*page* ix
Acknowledgements		xi

1 Foundations, federations, and finder charts — 1
  1.1 Star brightnesses — 2
  1.2 Absolute magnitude and distance modulus — 3
  1.3 Variable star nomenclature — 4
  1.4 Variable star classification — 8
  1.5 The *General Catalogue of Variable Stars* (*GCVS*) — 10
  1.6 Who wants your observations? — 11
  1.7 Finder charts and sequence charts — 13
  1.8 Light-curves and Julian Day Numbers — 16

2 Variables in vision — 20
  2.1 What type of telescope is best? — 20
  2.2 What size of telescope is best? — 23
  2.3 Eyepieces and fields of view — 29
  2.4 Vignetting — 31
  2.5 Binoculars — 36

3 Astrovariables reckoned — 39
  3.1 Preparations — 39
  3.2 Collimation — 42
  3.3 Finding your chosen variable — 52
  3.4 Making the magnitude estimate — 54
  3.5 Some difficulties and some remedies — 56

Contents

4 Photometry — 59
  4.1 Some basic principles of CCD astrocameras — 59
  4.2 The imaging area and resolution of a CCD camera when used on your telescope — 63
  4.3 CCD astrocameras in practice — 65
  4.4 Getting the focused image onto the CCD and keeping it there — 67
  4.5 Taking the picture — 70
  4.6 Calibration frames — 71
  4.7 Obtaining magnitude measures from a CCD image — 74
  4.8 Filters for photometry — 78
  4.9 Just the beginning — 80

5 Stars great and small — 81
  5.1 Our daytime star — 81
  5.2 Our stable Sun — 84
  5.3 Spectral lines — 88
  5.4 Stellar spectra — 92
  5.5 Information from spectra — 94
  5.6 Luminosity classes — 96
  5.7 The Hertzsprung–Russell diagram — 98

6 Variable beginnings — 100
  6.1 Single-star variables on the H-R diagram — 100
  6.2 Stellar nurseries within the interstellar medium — 100
  6.3 An unstable start in life — 105
  6.4 Stellar adolescence and the ZAMS — 108
  6.5 Stellar adulthood and stability — 111
  6.6 The fate of a low-mass star — 113
  6.7 The evolution of a star like the Sun — 117
  6.8 The evolution of a massive star — 119

7 Clockwork pulsators — 121
  7.1 A pulsating menagerie — 121
  7.2 The physics of stellar pulsation — 123
  7.3 CEP (Cepheid) and CEP(B) stars; DCEP (Classical Cepheid) and DCEPS stars; CW (W Virginis), CWA and CWB stars — 128
  7.4 RR (RR Lyrae), RR(B), RRAB, and RRC stars — 131

8 Less regular single-star variables — 133
  8.1 M (Mira) stars — 133
  8.2 SR (semi-regular variable); SRA; SRB; SRC; SRD; and SRS stars — 137
  8.3 A naked-eye hypergiant variable star — 141
  8.4 L (slow irregular variable); LB and LC stars — 143
  8.5 Other pulsating variable stars — 144

	8.6	RCB (R Coronae Borealis) stars	146
	8.7	GCAS (Gamma Cassiopeia) and B[e] stars	149
	8.8	Other single-star eruptive variables	150
	8.9	Rotating variable stars	151
9	**Eclipsing binary stars and novae**		**153**
	9.1	A matter of gravity	153
	9.2	Eclipsing binary stars	155
	9.3	Introduction to interacting stars	159
	9.4	N, NA and NB (classical novae), and NC stars	163
	9.5	NR stars (recurrent novae)	167
	9.6	Novae on the accompanying CD-ROM	168
	9.7	NL stars (nova-like variables)	169
	9.8	Nova hunting	170
10	**Cataclysmic and symbiotic systems**		**172**
	10.1	How to make a cataclysmic variable	172
	10.2	UG (U Geminorum) stars, aka dwarf novae	175
	10.3	Eclipsing dwarf novae	180
	10.4	Dwarf novae on the accompanying CD-ROM	183
	10.5	Polars, intermediate polars, and other cataclysmic subtypes	185
	10.6	ZAND (Z Andromedae) stars	186
	10.7	Intense X-ray sources	188
11	**The extra-galactic realm**		**190**
	11.1	Neutron stars	190
	11.2	Supernovae	191
	11.3	Supernovae on the accompanying CD-ROM	197
	11.4	Supernova hunting	197
	11.5	Black holes	199
	11.6	Hypernovae	200
	11.7	Quasars and active galaxies	203
	11.8	Cosmic chameleons	207
	11.9	Quasars and active galaxies on the accompanying CD-ROM	209
	Glossary		211
	Resources		217
	Index		221
	The accompanying CD-ROM		229

Preface

Stand outside to enjoy the glittering spectacle of a particularly clear night sky and you will probably get a false sense that the heavens are unchanging and serene. True, most of the stars visible do shine steadily but many do not. Some of them vary their brightnesses very slowly, taking years or centuries for any change to become apparent. Others that change do so faster, taking months or even just days. Still others can significantly vary their outputs in a matter of minutes. Some even flicker (in the real sense – not just the scintillation of their images as seen through our Earth's unsteady atmosphere) in timescales as short as seconds. Fortunately for us our Sun is one of the more constant of the 200 billion stars that inhabit our great Galaxy.

Actually, all stars must vary their outputs at some time – certainly during their births and deaths if not during other phases of their lives. Many stars are wrecked by colossal explosions and others are significantly changed by violent outbursts.

Variability is not the sole province of the stars. Galaxies, and particularly the objects lurking within their centres, can be subject to significant changes which involve energies of incredible proportions.

Astronomers both amateur and professional have long been following the behaviour of the variable-brightness objects in our Universe but it has fallen on amateurs to do most of the long-term monitoring. In recent years technical advances in the equipment available to amateur astronomers have pushed back the faintness limit and increased precision in the measurements. Consequently amateurs can now undertake work that was at one time the sole province of the professionals. You have a great opportunity to take part in this ongoing research yourself.

This book is intended to be a 'primer' – a guide for the interested amateur astronomer who is yet to become a specialist in the field of observing and monitoring variable stars and other objects of variable brightness.

Preface

In the first four chapters I cover the practicalities of observing and determining the brightnesses of the *astrovariable* – the term I have coined for all types of variable-brightness object in the heavens – at intervals which will allow their brightness changes to be studied.

Chapter 5 lays the basis for a study of a wide selection of astrovariables, this occupying the remaining chapters of this book. In these chapters I explain the reasons behind the brightness variations (as far as we presently understand them) set into the context of the wider field of astronomy and astrophysics. I think you will find it a fascinating story. Along the way we will make use of the considerable resources my friend and colleague, Nick James, has placed on the CD-ROM which accompanies this book.

I hope that you will enjoy reading this book. Most of all, though, I hope that you will go out and use whatever equipment you can assemble to begin observing the variable heavens for yourself. If you do, I hope that the information and resources in this book and the accompanying CD-ROM will help you along the way. Good luck – and good observing!

Acknowledgements

I have the following friends, colleagues, organisations, and companies to thank for allowing me to use illustrative materials in this book and on the accompanying CD-ROM:

Martin Mobberley; Tom Boles; Terry Platt; Starlight Xpress Ltd; W. J. Worraker; Denis Buczynski; Mark Armstrong; the Royal Greenwich Observatory (RGO); the National Aeronautics and Space Administration (NASA); The Hubble Heritage Team and The Space Telescope Science Institute (STScI).

I also offer my special thanks to Dr Nick Hewitt, Roger Pickard, and the Council of the British Astronomical Association (BAA) for their kindness in allowing me to reproduce materials from the archives of the Variable Star Section of the BAA in the pages of this book and, most especially, in the accompanying CD-ROM. Guy Hurst of *The Astronomer Magazine* (TA) group has also kindly given his permission for the group's archives to be reproduced in this book and, particularly, on the accompanying CD-ROM. My special thanks also extend to N. N. Samus and O. V. Durlevich, of the Sternberg Astronomical Institute of Moscow, for allowing me to reproduce materials from the *General Catalogue of Variable Stars* (GCVS) Research Group.

I must also thank Nick James, who has allowed me to use some of his images in the main text and in the accompanying CD-ROM, but particularly for his production of the CD-ROM which accompanies this book.

Finally I must thank Jacqueline A. Garget and all the staff at Cambridge University Press for their stirling work in producing this book. To all the above I offer my very grateful thanks.

Figure 1.1 The constellation of Orion, photographed by the author. Of the main stars that form the familiar outline of this constellation, the upper-left one is the red giant star Betelgeuse. This semi-regular variable star is one of about thirty whose brightness variations can be followed with the naked eye.

Chapter 1
Foundations, federations, and finder charts

In this book we will examine the various reasons why some celestial bodies vary their luminosities, in addition to tackling the practicalities of observing them and following their brightness changes. The names of objects and phenomena such as eclipsing variable stars, pulsating variable stars, symbiotic stars, eruptive variable stars, cataclysmic variable stars, novae, supernovae, hypernovae, X-ray bursters, γ-ray bursters, Active Galactic Nuclei, Seyfert galaxies, BL Lacertae objects, quasars, and more are our main fare. I propose the generic term *astrovariables* for them.

You might not have access to expensive equipment and only have a little of your time and the use of your eyes which you can spare. Even so, you can still make a contribution. If you doubt this, take a look at Figure 1.1 which shows the constellation of Orion. Of the main stars, the upper-left one is the red giant Betelgeuse. It is a variable and, using the techniques described later in this book, its variations of brightness can be followed by making estimates with no equipment other than the unaided eye. There are other examples.

Still, if you have or can obtain some cheap equipment then all the better because this widens the scope enormously. With very limited resources you will have a wide enough choice of objects to follow to potentially fill many lifetimes of study! If you can increase your equipment budget to a few thousand dollars then you can emulate professional astronomers and produce cutting-edge research work. Let me emphasise, though, that observing astrovariables is not the sole province of the wealthy with loads of time on their hands. You can experience a lifetime's fascination and pleasure by observing astrovariables with very limited resources – and produce scientifically valuable work while you are doing so!

Later there will be more about the astrophysics of astrovariables, and how to observe examples of each type of them. In order to use the space in the rest of this book efficiently, though, I ought to use this preliminary chapter to cover a few of the fundamentals it will be useful to have to hand. Let us begin by defining the brightness scale which is the very foundation stone of our work.

1.1 Star brightnesses

The *apparent visual magnitude* of a star is a measure of how bright it *appears* to be in our sky. The magnitude scale can cause confusion to the uninitiated because the larger positive number actually corresponds to the dimmer star.

The lovely steely-blue coloured Vega (the brightest star in the constellation of Lyra) is defined to have an apparent visual magnitude of $0^m.0$. There are four stars which appear brighter than Vega and so they are given negative apparent magnitudes. The brightest of these is the brilliant white Sirius, which has a magnitude of $-1^m.5$.

What about the other detectors we can use in astronomy? The wavelength (colour) to which they are most sensitive is a little different from detector to detector – and often very different to the response of the human eye. Star brightnesses measured with different detectors come out a little different because of this, since stars also differ in colour. This is why we make the distinction 'visual' in apparent visual magnitude, the term for a star's brightness as seen by the human eye. There is more about this in Chapter 4. In this book please assume that I am referring to visual magnitudes (as seen by the human eye, or a device which mimics the same response) unless indicated otherwise. Hence 'Sirius has a magnitude of $-1^m.5$' refers to its visual magnitude.

The magnitude scale is not linear (equal steps for equal brightness changes) but is instead based on ratios, with each magnitude difference corresponding to a brightness difference of 2.5 times. A difference of 5 magnitudes corresponds to a brightness difference of $2.5 \times 2.5 \times 2.5 \times 2.5 \times 2.5$ times, or 100 times (in truth the ratio is more accurately 2.512).The reason for this is that the eye appreciates brightness differences in terms of ratios and so the empirical magnitude figures that were originated by the astronomers of long ago corresponded to ratios of brightness of about this figure of 2.5.

Mathematicians define such a scale, where equal steps represent a change by a constant multiplication factor, to be a *logarithmic* scale. Looking at this from a mathematician's point of view, if we say that a number N is equivalent to another number of the form a^x (for instance $100 = 10^2$), then we can write a relationship between these numbers in terms of a logarithm. The relationship is $\log_a N = x$ (for example $\log_{10} 100 = 2$). Logarithms are an artificial construct but they do lend themselves to conveniently representing and manipulating numbers, such as those we meet in our work of measuring the brightnesses of astrovariables.

Let me restate the relationship, known as the *log identity*, that defines a logarithm:

if $N = a^x$, then $\log_a N = x$

The figure a is known as the *base* of the logarithm. In our work we will only be interested in logarithms of base 10 ($a = 10$). For this special case we do not need to bother to write $\log_{10}$ each time. Instead we can write Log (note the capital L).

Here is the basis of the stellar magnitude scale:

$$m = -2.5 \operatorname{Log} I$$

where I is the apparent luminosity of the star, in relative units, and m is its resulting apparent magnitude.

The difference in apparent visual magnitude between one star and another, Δm, is then given by:

$$\Delta m = 2.5 \operatorname{Log}\left(\frac{I}{I'}\right)$$

where I and I' are the relative brightnesses of the stars, I being the intensity of the brighter star (which, remember, also has the *lower*, or *more negative*, magnitude number). Can you see what the magnitude difference is between two stars where the brighter star has 1000 times the luminosity of the other? The answer is 7.5 magnitudes.

On the very best nights, if you have very keen eyesight and are sited well away from any sources of 'light pollution', you could expect to see stars down to magnitude $6^m.5$. Of course an observer with exceptionally acute vision and access to a superior site might do better – a few observers have claimed to see stars as faint as eighth magnitude from some mountaintop locations using nothing but their unaided eyes. More prosaically, from my garden in a semi-rural English village I rarely get evenings clear enough to show stars fainter than about magnitude $5^m.0$ without using optical aid.

1.2 Absolute magnitude and distance modulus

Since stars display a great range in their actual luminosities, the apparent magnitude of a star is by no means a reliable guide to its distance. We measure the distances of the nearby stars by the method of *trigonometrical parallax*. This is where a star's apparent position is measured with respect to several others. As the Earth moves around the Sun, so a nearby star will apparently shift its position with respect to the more distant stars. The extremes of position occur six months apart, since this is the period over which the Earth travels halfway round its orbit, providing the baseline for our changing viewpoint. Half the total angular shift of the star is the parallax.

So far no star has been found with a parallax greater than 1 arcsecond. A 1 arcsecond parallax would mean that the star was a distance of 206 265 AU from us. This is 30 million million kilometres, a distance we prefer to call 1 *parsec*. A parsec is 3.26 light years. The number of parsecs is found by taking the reciprocal of the number of arc seconds of parallax. For instance, a star that has a parallax of 0.5 arcsecond is 2 parsecs, or 60 million million kilometres from Earth.

Foundations, federations, and finder charts

The difficulty of measuring tiny angular movements using our Earth-based telescopes had put a limit of about 100 parsecs on the distance for which we could use parallax. However, the *Hipparcos* satellite launched into Earth orbit by the European Space Agency observed and precisely measured the positions, brightnesses, colours, and parallaxes of over a hundred thousand stars with milliarcsecond accuracy during the years 1989–1993. The distances of the stars within a hundred parsecs, or so, are now known with an accuracy of around 1 per cent and the range at which parallaxes are still useful extends about ten times as far. The data from *Hipparcos* are still having an impact on many branches of research and yet are to be bettered by many orders of magnitude by the proposed *GAIA (Global Astrometric Interferometer for Astrophysics)* probe, presently slated for launch sometime around 2012.

If we know how far away a star is and we measure its apparent brightness, then we can find its real luminosity. This is often expressed as its *absolute magnitude*. The absolute magnitude of a star is equal to the apparent magnitude it would have if it was set at a standard distance of 10 parsecs from Earth.

The Sun's apparent magnitude is $-26^m.7$, but its absolute magnitude is $4^m.8$, so it would appear rather insignificant if it were placed at the standard distance of 10 parsecs away. Absolute magnitude is denoted by M to distinguish it from apparent magnitude, m.

The quantity $(m - M)$ is useful, as it fixes the distance of a given star. Alternatively, if the apparent magnitude and distance of the star are measured, then its true luminosity can be found. The quantity $(m - M)$ is known as the *distance modulus* of the star. The equation that relates m, M, and the distance of the star in parsecs, d, is:

$$(m - M) = (5 \operatorname{Log} d) - 5$$

1.3 Variable star nomenclature

Since there are far too many stars (and other astronomical objects) to have them all given proper names, the next best thing is to use a scheme based on the genitives of the names of the host constellations. One ingredient of this scheme (although only for the brightest stars) is an assigned Greek letter, originally devised as expressing the rank order of brightness of the star in the constellation (though certainly discrepancies exist in the stars as we see them today). Consequently Vega, the brightest star in the constellation of Lyra, is also known as α Lyrae. This can be abbreviated to α Lyr.

You will probably already be very familiar with all this but I thought it would be useful to present the constellations' names with their genitive forms and their abbreviations all together here in Table 1.1. These genitive forms themselves recur time and time again in astronomy and are the basis for the main schemes of naming astrovariables. Table 1.2 provides a listing of the Greek alphabet, also for your convenience.

1.3 Variable star nomenclature

Table 1.1 *Constellation genitive forms and abbreviations*

Constellation	English name	Genitive	Abbreviation
Andromeda	Andromeda	Andromedae	And
Antlia	The Airpump	Antliae	Ant
Apus	The Bird of Paradise (or the Bee)	Apodis	Aps
Aquarius	The Water-bearer	Aquarii	Aqr
Aquila	The Eagle	Aquilae	Aql
Ara	The Altar	Arae	Ara
Aries	The Ram	Arietis	Ari
Auriga	The Charioteer	Aurigae	Aur
Boötes	The Herdsman	Boötis	Boo
Caelum	The Sculptor's Tools	Caeli	Cae
Camelopardalis	The Giraffe	Camelopardalis	Cam
Cancer	The Crab	Cancri	Cnc
Canes Venatici	The Hunting Dogs	Canum Venaticorum	CVn
Canis Major	The Great Dog	Canis Majoris	CMa
Canis Minor	The Little Dog	Canis Minoris	CMi
Capricornus	The Sea-goat	Capricorni	Cap
Carina	The Keel (of the ship Argo)	Carinae	Car
Cassiopeia	Cassiopeia	Cassiopeiae	Cas
Centaurus	The Centaur	Centauri	Cen
Cepheus	Cepheus	Cephei	Cep
Cetus	The Whale	Ceti	Cet
Chameleon	The Chameleon	Chameleontis	Cha
Circinus	The Compass	Circini	Cir
Columba	The Dove	Columbae	Col
Coma Berenices	Berenice's Hair	Comae Berenices	Com
Corona Austrinus	The Southern Crown	Coronae Austrina	CrA
Corona Borealis	The Northern Crown	Coronae Borealis	CrB
Corvus	The Crow	Corvi	CrV
Crater	The Cup	Crateris	Crt
Crux	The Southern Cross	Crucis	Cru
Cygnus	The Swan	Cygni	Cyg
Delphinus	The Dolphin	Delphini	Del
Dorado	The Swordfish	Doradus	Dor
Draco	The Dragon	Draconis	Dra
Equuleus	The Foal	Equulei	Equ
Eridanus	The River Eridanus	Eridani	Eri
Fornax	The Furnace	Fornacis	For
Gemini	The Twins	Geminorum	Gem
Grus	The Crane	Gruis	Gru
Hercules	Hercules	Herculis	Her
Horologium	The Clock	Horologii	Hor
Hydra	The Sea-serpent	Hydrae	Hya
Hydrus	The Watersnake (or Small Sea-serpent)	Hydri	Hyi
Indus	The Indian	Indi	Ind

(*cont.*)

Table 1.1 (cont.)

Lacerta	The Lizard	Lacertae	Lac
Leo	The Lion	Leonis	Leo
Leo Minor	The Little Lion	Leonis Minoris	Lmi
Lepus	The Hare	Leporis	Lep
Libra	The Scales	Librae	Lib
Lupus	The Wolf	Lupi	Lup
Lynx	The Lynx	Lyncis	Lyn
Lyra	The Lyre	Lyrae	Lyr
Mensa	Table Mountain	Mensae	Men
Microscopium	The Microscope	Microscopii	Mic
Monoceros	The Unicorn	Monocerotis	Mon
Musca Australis	The Southern Fly	Muscae	Mus
Norma	The Rule	Normae	Nor
Octans	The Octant	Octantis	Oct
Ophiuchus	The Serpent-bearer	Ophuichi	Oph
Orion	Orion (the Hunter)	Orionis	Ori
Pavo	The Peacock	Pavonis	Pav
Pegasus	The Winged Horse	Pegasi	Peg
Perseus	Perseus	Persei	Per
Phoenix	The Phoenix	Phoenicis	Phe
Pictor	The Painter	Pictoris	Pic
Pisces	The Fishes	Piscium	Psc
Piscis Austrinus	The Southern Fish	Piscis Austrini	PsA
Puppis	The Poop-deck (of the ship Argo)	Puppis	Pup
Pyxis	The Mariner's Compass	Pyxidis	Pyx
Reticulum	The Net	Reticuli	Ret
Sagitta	The Arrow	Sagittae	Sge
Sagittarius	The Archer	Sagittarii	Sgr
Scorpius	The Scorpion	Scorpii	Sco
Sculptor	The Sculptor	Sculptoris	ScI
Scutum	The Shield	Scuti	Sct
Serpens Caput	The Serpent's Head	Serpentis	Ser
Serpens Cauda	The Serpent's Tail	Serpentis	Ser
Sextans	The Sextant	Sextantis	Sex
Taurus	The Bull	Tauri	Tau
Telescopium	The Telescope	Telescopii	Tel
Triangulum	The Triangle	Trianguli	Tri
Triangulum Australe	The Southern Triangle	Trianguli Australis	TrA
Tucana	The Toucan	Tucanae	Tuc
Ursa Major	The Great Bear	Ursae Majoris	UMa
Ursa Minor	The Little Bear	Ursae Minoris	UMi
Vela	The Sails (of the ship Argo)	Velorum	Vel
Virgo	The Virgin	Virginis	Vir
Volans	The Flying Fish	Volantis	Vol
Vulpecula	The Fox	Vulpeculae	Vul

1.3 Variable star nomenclature

Table 1.2 *The Greek alphabet*

Alpha	α	Nu	ν
Beta	β	Xi	χ
Gamma	γ	Omicron	o
Delta	δ	Pi	π
Epsilon	ε	Rho	ρ
Zeta	ξ	Sigma	σ
Eta	η	Tau	τ
Theta	θ	Upsilon	υ
Iota	ι	Phi	φ
Kappa	κ	Chi	χ
Lambda	λ	Psi	ψ
Mu	μ	Omega	ω

In the nineteenth century Friedrich Argelander originated the scheme we still use today for naming variable stars. In this scheme the first discovered variable star in a constellation was given the letter R followed by the genitive form of the constellation name. For instance the first variable star discovered in Cygnus was named R Cygni. It still is known by this name. The second, third, and fourth stars discovered in Cygnus are S Cygni, T Cygni and U Cygni. The ninth variable to be discovered in Cygnus is, of course, Z Cygni. When a tenth variable star was discovered in a given constellation a double letter prefix was used, for instance RR Cygni. The scheme was continued with RS Cygni, then RT Cygni, RU Cygni, and so on.

After RZ Cygni the sequence begins again with SS Cygni (*the second letter must not be earlier in the alphabet than the first – so SR Cygni is NOT permitted*), ST Cygni and onwards to SZ Cygni. After SZ Cygni comes TT Cygni, then TU Cygni and ... you get the idea.

Eventually all the designations up to ZZ were used and so astronomers reverted to using double letters in the first part of the alphabet: AA to AZ, then BB to BZ, then CC to CZ, and so on. The letter J was never used in case it might be confused with the letter I when written.

The foregoing schemes allowed the designation of 334 stars in any given constellation, the last being QZ (remember, RR to ZZ were already used up). Eventually still further variables stars were discovered. For these a V is used followed by a number and the constellation name. For instance, the 335th variable star in Orion is V335 Orionis, while the next one discovered is V336 Orionis.

Other schemes for naming stars are also in vogue. For instance, *Harvard Designations* (*HD*) are also commonly used. In this scheme the star is given a six-digit number which represents its co-ordinates for epoch 1900. The first two digits give the number of hours of right ascension (00 to 24) and the second two give the remaining number of minutes (00 to 59). The final pair of digits give the declination of the star (and these are in italics for negative declinations).

As an example, the Harvard Designation for the star R Cygni is HD 193408 because its co-ordinates were $\alpha = 19^h\ 34^m$, $\delta = +08°$ on 1 January 1900. To take another example, the Harvard Designation of the star R Centauri is HD 140959 because its co-ordinates on 1 January 1900 were $14^h\ 09^m$, $-59°$.

In the cases where you know both the Harvard and Argelander designations of variable stars, it is a good idea to give both when reporting observations as this will minimise the risk of misidentifying the star to which your observation corresponds.

For example, writing 193408 R Cygni confirms to the recipient of your observations which star your observed magnitude corresponds to. If you referred to a particular star as 200938 R Cygni he/she would be alerted that there is a problem. He/she could ask you to check and you would find that you should have written the star as 200938 RS Cygni. Without the Harvard Designation as a check your value of the observed brightness of the star RS Cygni would have been assigned to the star R Cygni!

There are many more star catalogues I could mention. In due time you will undoubtedly encounter stars with designations beginning HIP (from the catalogue created from the *Hipparcos* database) and SAO (the important catalogue issued by the Smithsonian Astrophysical Observatory in 1966), along with many others but I will stop here. The Argelander and Harvard Designations are the main schemes of use to us and it would only use up precious space in this book, and maybe even confuse matters, to delve into the others. This is enough to get you underway – and it is my intention throughout this book to present you with enough useful information to get you started, while not including so much as to mire you in minutia.

1.4 Variable star classification

I wish I could say that the classification scheme for variable stars was easy, simple, and straightforward. Unfortunately I cannot. When we get down to the really fine detail I cannot even say that it is entirely permanent! Please do not allow yourself to become too bogged down with how variable stars are classified. After all, it is the stars themselves that really matter. I think it will be of the greatest help to you at this juncture if I present a broad picture of the way variable stars are classified and then introduce specific cases as we come to them in the course of this book.

The classification of variable stars can be thought of as existing in layers. In the first layer we can divide all variable stars into just two types: *intrinsic variable stars* if their brightness varies due to some internal cause and *extrinsic variable stars* if it varies due to some external agency

In the next layer we can divide the same stars this time into seven groups: *eruptive variable stars*; *pulsating variable stars*; *rotating variable stars*; *cataclysmic variable stars*, *eclipsing binary systems*; *optically variable X-ray sources*; and '*other miscellaneous variables*'. I choose these seven groups but some authorities would

1.4 Variable star classification

insist that the entire population of variable stars ought to be divided up in different ways. Just as one instance *symbiotic stars* could be considered as distinct from cataclysmic variable stars. I could go on but I am sure you get the idea. I think that the seven groups I have already stated are sufficient to cover all bases, each group being considered a 'broad church' itself containing diversity in its members.

There is a particular reason why I am promoting these seven categories as standards we should accept – and that is because it is the scheme adopted by the very important *General Catalogue of Variable Stars (GCVS) Research Group*, based at the Sternberg Astronomical Institute of Moscow, in Russia. This scheme has been developed by the expert members of the group, incorporating advice from a number of specialists. I have more to say about this group in the next section, suffice it to say here that I recommend you adopt this group's classifications when starting out in your own researches.

You must, though, be prepared for variations in the way different authors and authorities treat and classify astrovariables. For instance, they can also be divided up into groups based on the way their brightnesses vary with time. So some authorities speak of *irregular variable stars* and *semi-regular variable stars*, *long-period variable stars*, *novae*, *dwarf novae* and *recurrent novae*, *and flare stars*.

So, all variable stars can be classified into groups according to either scheme. The third layer of classification involves grouping stars into those that show similar behaviour, tempered by our understanding of their physical nature and the processes in operation causing their variability. In other words, we group stars together according to their characteristics. Each of these groupings we give a designation based on a chosen star that exemplifies the group.

For example, one class of very young variable stars are known as *FU Orionis stars*. All FU Orionis stars behave like and, as far as we can at present deduce, are physically very much like FU Orionis itself. These stars undergo irregular surges in brightness and so would be considered as members of the broader set of stars known as eruptive variables. Of course that also makes them irregular variables under the alternative scheme.

Again, I wish to promote the classifications issued by the GCVS group as the standards we should adopt. The GCVS variable groupings are often written as abbreviations. Hence FU Orionis stars are denoted as FU.

At this point I want to introduce you to the CD-ROM that accompanies this book. It is full of resources to help you. If all the files were printed out, the volume of material on the CD-ROM would fill this book several times over – which is why it is presented on the CD-ROM and not in these printed pages!

Once it is started on your computer you just click on the 'Index' icon and you will see a menu. One of the items listed under 'Miscellaneous' is a text file called 'GCVS variability types'. Click on this and you will see displayed a partially updated document first written by Dr Nikolai N. Samus and Dr O. V. Durlevich in 1998 (and reproduced by special permission). You will see it contains a full and detailed listing, with explanations, of the classification scheme adopted by the GCVS Research Group. The listing includes the abbreviations for each type.

Please do be aware that even this set of designations is not itself fixed and constant. At present there are over ninety groupings named after 'prototype' stars but the number keeps on growing as smaller and smaller distinctions are being recognised. There are even subdivisions to some of these.

Please also do be aware that some of the groupings are known by alternative names, especially by other authorities. For example, the stars that used to be known as β Canis Majoris stars are now fashionably called β Cephei stars and BL Boo stars can also be referred to as 'anomalous Cepheids'. It is also true that as a variable star's light variations and other physical parameters become increasingly refined with continued observation this sometimes leads to the star being reclassified from one type of variable to another.

Personally, I think that the classifying of variable stars has evolved into something of a headache – and this is also a reason why the classifications of such a major and important body as the GCVS Research Group ought to be adopted as standard. All I can say is don't have bad dreams about it yourself. Watch for inconsistencies between authors and be especially careful when you are reading older literature on this subject.

I will, in the course of this book, introduce the main GCVS groupings and subtypes of variable stars as we come to them. However, I have no intention of even trying to be complete in this respect. This book is to help you get started in your study of variable stars – but most particularly concerns how to actually observe them. Learning all the types of variable stars and their particular characteristics is something that you can take your time doing. It may be that eventually you will develop a special interest in one class of variable stars at the expense of all others. You might then not be interested in the other types at all. However, please do remember that the GCVS classification document is on the CD-ROM for you to refer to whenever you need it.

Incidentally, the end part of the 'GCVS variability types' document also contains a listing of the numbers of objects (of the original 35 148 objects listed in 1998) that fall into each designation. You will see that a great many of the specific variable types are extremely thinly represented, some having just one member!

1.5 The *General Catalogue of Variable Stars* (*GCVS*)

As well as covering practical matters involved with making observations, this book includes introductions to the various types of variable object and how they fit into the grand scheme of things. Along the way I consider particular examples. However, I can only cover so much in the available space and you certainly will not want to limit yourself to studying just the examples I give. You will find further examples of astrovariables to study in various magazines and journals and you should ask for advice on the current programmes carried out by whatever astronomical society/association you belong to from the relevant Director/Co-ordinator.

However, there is one particular resource that you should be aware of, whether you are a newcomer or a seasoned practitioner. That resource is the *General Catalogue of Variable Stars*, generally just known as the *GCVS*. I have already referred to the GCVS classification scheme (reproduced on the accompanying CD-ROM) but the GCVS Research Group offer a vast resource of information and research material. In particular they have listed nearly 40 000 designated astrovariables in five main catalogues with additional supplements such as *Name-Lists of Variable Stars* (NL), *The New Catalogue of Suspected Variables* (NSV), and others.

You can access this material yourself on the Internet. Either use a search engine such as *Google* (typing 'General Catalogue of Variable Stars'), or you can access GCVS Research Group homepage directly at:

http://www.sai.msu.sv/groups/cluster/gcvs/gcvs/

One of many particularly useful functions available on the homepage is the *GCVS Query form*. Click on that and you will be instructed to input the GCVS designation of the astrovariable. Out will pop a short description of the object's characteristics complete with various relevant cross-references. You can also cross-reference the named GCVS star to other catalogues (ADS, HIP, SAO, and many others.) The GCVS website is a fabulous resource and when you feel ready to begin seriously researching astrovariables on your own behalf it is one I heartily recommend to you.

1.6 Who wants your observations?

There are no laws to say that you cannot spend the rest of your life working at your astronomical observations in total isolation. You could keep all your results to yourself. No one will send the police round to arrest you for your isolationism. However, it would still be a crime. There is so much pleasure and camaraderie you can gain by joining an astronomical society or association and so much personal satisfaction in pooling your observations with those of other people that not to do so would be folly, and a terrible waste of the results of your work.

Further, astrovariable observing is a field where the results of one person's work can be of limited value when taken in isolation, especially so where the making of visual estimates is concerned. However, each contributor's observations takes on a greatly enhanced value when combined with the observations of other people.

If you want the greatest pleasure and satisfaction from your observing and want your results to have their greatest value, then join the observing group of a national or provincial society and submit your observations to a central co-ordinator (often known as an observing section 'Director'). You will find that the co-ordinator will be very grateful for whatever results you can send in and will be on hand to advise you on selecting objects for study and the various practical matters involved in making and submitting your observations. You will also get feedback from your observations in the form of reports/circulars, articles in journals, etc. which will bring to life for you the observing work you undertake.

The first step is to contact the Secretary of the Association/Society for details about membership and the various observing sections the society possesses (with contact addresses of the Directors/Co-ordinators). Following is a limited list of various national associations/societies which have sections devoted to variable star observing, or are entirely devoted to this work:

In the UK the premier national amateur astronomical society is the *British Astronomical Association* (*BAA*). The address is:

Burlington House, Piccadilly, London W1J 0DU, England.
Telephone: 020-7734 4145
Fax: 020-7439 4629
Home page: http://www.britastro.org/

Of particular note to us is the Variable Star Section of the BAA (known as the BAAVSS). Its Web page contains a vast resource of additional information, such as details of ongoing research programmes, a list of all objects for which the BAAVSS has a database, the most up-to-date charts, etc.:

http://www.britastro.org/vss

There is a separate group operating from the UK but with an international membership which is set up with separate observing sections like the BAA, which has strong links with the BAA, and which has particular strong professional–amateur links and is internationally respected. This is: The Astronomer Magazine (TA). TA exists mainly as a clearing house for observations and it is a very valuable resource for up-to-date information on observations and discoveries in variable star and other aspects of observational astronomy. It has no one main address but the Secretary and each of the leaders of the observing sections are best contacted via the addresses, telephone numbers, and/or email addresses given on the 'contact details' link on the TA homepage:

http://theastronomer.org/index.html

The United States of America has an observing society which specialises in variable stars: the American Association of Variable Star Observers (AAVSO). The address is:

25, Birch Street, Cambridge, MA 02138, USA.
Home page: http://www.aavso.org/

In Japan there is The Variable Star Network (VSNET), of Kyoto University. The address is:

Kitashirakawa-Oiwake-cho, Saykyo-ku, Kyoto 606–8502, Japan.
Home page: http://www.kusastro.kyoto-u.ac.pp/vsnet

An association in Germany devoted solely to variable stars is the Bundesdeutsche Arbeitsgemeinshaft für Veränderliche Sterne (BAV). The address is:

Munsterdamm 90, 12169 Berlin, Germany.
Home page: http://thola.de/bav.html

The equivalent association in France is the Association Française des Observateurs d'Etoiles Variables (AFOEV). The address is:

Observatoire Astronomique de Strasbourg, 11 rue de l'Université,
67000 Strasbourg.
Home page: http://cdsweb.u-stasbg.fr/afoev/

Observers of the southern skies are served by the Astronomical Society of South Australia (ASSA), whose Secretary can be contacted at:

GPO Box 199, Adelaide, SA 5001, Australia.
Home page: http://www.assa.org.au/info/

– and by the Royal Astronomical Society of New Zealand (RASNZ), whose Secretary can be contacted at:

PO Box 3181, Wellington, New Zealand.
Home page: http://www.rasnz.org.nz/

1.7 Finder charts and sequence charts

How do you track down a particular variable star? Once you have found it, how do you go about determining its brightness? This is a relatively straightforward affair for naked-eye stars, especially if you have a star chart or atlas but for the fainter ones (particularly for those visible only with optical aid) you will need a *finder chart* to help you.

Part of the support that you can expect from joining a society/association with an active variable star observers' section is the facility to obtain finder charts. Of course, you will need at least one finder chart (separate charts for binocular and telescope fields, where appropriate) for each astrovariable. These can usually be obtained at very small cost, per chart, on request. However, the cost of obtaining a very large selection of charts would soon mount up and the filing system needed for them might be problematical. If you are a complete beginner, there is also the headache of deciding which astrovariables you should pursue (and so which charts to purchase), although the association Director/Co-ordinator will undoubtedly offer advice on this score.

I can also offer you some practical help. On the CD-ROM that accompanies this book you will find finder charts for approximately 600 astrovariables, most of these being for variable stars. My friend and colleague, Nick James, has prepared this valuable resource from the archives of TA and the BAAVSS. These charts are reproduced courtesy of these organisations and with their special permission. The chances are fairly good that almost any astrovariable you choose to observe (especially while you are in your 'apprenticeship phase') will have its finder chart on this CD-ROM. At the very least this resource gives your collection of finder charts a tremendous start at no extra expense to yourself!

If finding the charts for a particular object amid the long lists of them on the CD-ROM seems daunting, remember the CD-ROM works with your web-browser and so you can use its 'find' function to locate the entry.

For instance when running the CD-ROM with 'Microsoft Internet Explorer Version 4.0' on my geriatric computer, I can view the chosen menu of charts (let us say the TA charts) and then click on 'Edit'. I click on the option 'Find' and type in the name of my selected file (e.g. 'OS And') in the waiting box, then click 'Find Next' and the computer immediately shows me the relevant part of the chart menu showing the selected file (e.g. 'OS And') highlighted. After locating the relevant file you can then go ahead and view and/or print out the chart you have selected in the usual way.

The TA charts can come in any of three versions. 'A' charts are wide field charts with naked-eye orientation, designed for the observer who uses binoculars. 'B' charts are narrow field and orientated for a standard astronomical telescope. 'C' charts are designed for deep (i.e. faint object) telescopic and CCD use. When you examine the TA charts on the CD-ROM you will find that many objects are represented by two charts, for instance 'B' and 'C', and some by all three versions. By way of an example, Figure 1.2(a), (b), and (c) shows the TA charts for OS Andromedae in all three versions.

The BAA finder charts are denoted as 'A', 'B', 'C', etc. in order of increasing scale (progressively narrower field of view), though in most cases just one scale of chart, denoted 'A', exists for each named astrovariable. You will probably have to resize these charts for your own use and to print them out.

You will notice that the charts also show the magnitudes of several of the stars surrounding the astrovariable. It is from the magnitudes of these stars that you will be able to determine the magnitude of the variable. So, these finder charts are also *sequence charts*. Most of the charts on the CD-ROM are sequence charts as well as being finder charts.

The co-ordinator of whichever observing group you belong to will normally require you to use the charts from a specific organisation – most usually the BAA, TA, or the AAVSO but there are others. This is because there is some discrepancy between the magnitudes of some comparison stars. It is obviously best if all the observers in an organisation are using the same comparison sequences.

The AAVSO charts can be accessed via the website of the organisation. The website houses over 3000 charts divided into *constellation finder charts/*

1.7 Finder charts and sequence charts

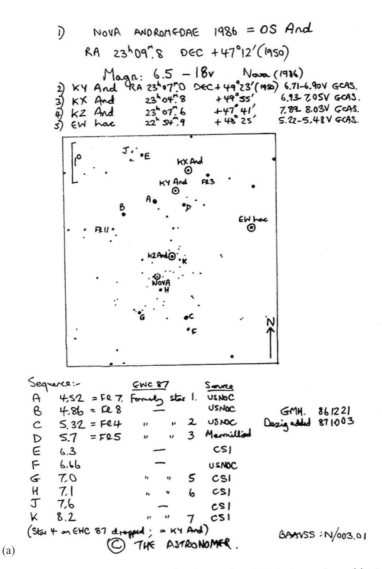

Figure 1.2 Finder (and sequence) charts for OS Andromedae: (a) A-scale chart; (b) B-scale chart (see page 16); (c) C-scale chart (see page 17). Courtesy The Astronomer Magazine (TA) and the British Astronomical Association Variable Star Section (BAAVSS).

educational charts (very wide field plots), *standard charts* (the main ones to use), *preliminary charts* (for newer objects awaiting establishment of the comparison star sequences), *special purpose charts* (generally for advanced observers with photometric equipment), and *reversed charts* (mirror-reversed and inverted charts for use with modern equipment, for instance a Schmidt–Cassegrain telescope fitted with a star diagonal).

Unlike the TA and BAA charts, those of the AAVSO have the comparison magnitudes marked next to the stars on the pictured field of view. However, the

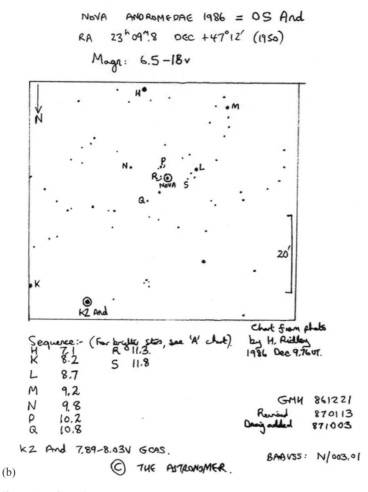

Figure 1.2 (cont.)

decimal points are not included in case they are confused with stars. You might see a number such as 89 against a certain star meaning that particular star is of magnitude $8^m.9$.

Of course, you can examine the charts on your computer monitor but you will normally want to print out your chosen chart or charts for use at your telescope.

1.8 Light-curves and Julian Day numbers

Another resource on the accompanying CD-ROM is a set of graphs of magnitude vs. time, known as *light-curves*, for more than 500 astrovariables. These are from the BAAVSS archives and are reproduced by special permission. You can use the 'find' function of your web-browser to locate specific examples, in the same

1.8 Light-curves and Julian Day numbers

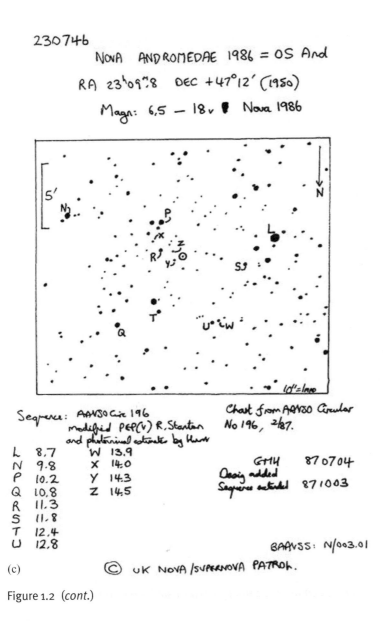

Figure 1.2 (cont.)

way as you can find specific finder charts. Figure 1.3(a), (b), and (c) shows just three examples and several more are reproduced later in this book. Light-curves are a very visual way of showing how the brightness of an astrovariable varies with time.

When you have collected enough data points (magnitude determinations) for any chosen astrovariable you can draw up your own light-curves. This obviously works best for variables which show either slower, long-term, brightness variations – or those which change very rapidly, say in the course of a few hours so you can monitor the changes in a single observing session. In general, though,

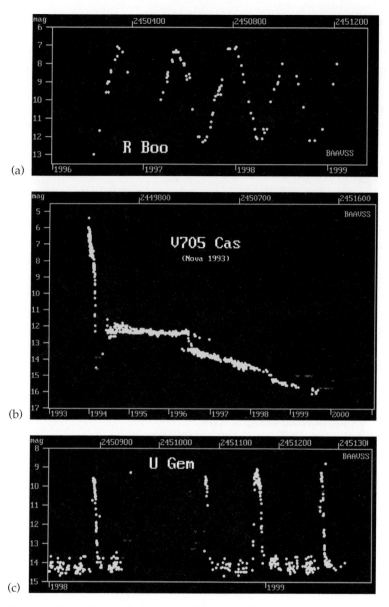

Figure 1.3 Variable star light-curves: (a) R Boo; (b) V705 Cas – a nova which was seen to erupt in 1993; (c) U Gem. The gap in the light-curve in (c) results from U Gem appearing too close to the Sun to observe for part of the year. The numbers along the bottom of the figure refer to dates in years. The figures along the top refer to dates in Julian Day (JD) number. Courtesy BAAVSS.

1.8 Light-curves and Julian Day numbers

the best light-curves are generated from the pooling of the results from many observers in order to get the best possible coverage and time resolution.

Going beyond light-curves, one can make other investigations from the results, for example determining the period of the brightness variations and looking for multiple superimposed periods of brightness variation. However, that does take us into rigorous mathematical (or computing) techniques which are beyond the remit of this book. Also, that sort of analysis is most definitely best carried out by the person who has access to the results of many contributors. A mere handful of data points will not be sufficient to analyse an astrovariable's behaviour in any meaningful way.

You may notice in Figure 1.3 (and this is also true of a large fraction of the light-curves on the CD-ROM), that the units given on the time axis are 'JD'. JD stands for *Julian Day* (or *Julian Date*). On this artificial system Julian Day 1 began at noon (Universal Time) on 4713 January 1 BC. Thus each new Julian Day begins at noon. Noon on 2003 January 1 marks Julian Day number (JD) 2452641.0. Midnight on the same day is JD 2452641.5.

The reason for expressing times as JD numbers is to make the subsequent analysis of an astrovariable's behaviour easier. It avoids having to reckon with years, months, and the differing numbers of days in any one month (30 and 31, also February's 28 but 29 in a leap year). The system was devised with the 'day' beginning at noon in order to avoid a day change occurring during a night-time observing session.

You will find Julian Day numbers given in almost any astronomical ephemeris – but do remember the Julian Day begins at $12^h\ 00^m$ UT and you have to use your calculator to find the fraction of the 24 hour period to the time of your observation that has elapsed since then in order to fill in the number after the decimal point.

Now we have covered some of the fundamentals, we should next turn our attention to the 'tools of the trade': the optical equipment we intend to use for our observing. That is the subject of the following chapter.

Chapter 2
Variables in vision

In this chapter we will consider the main instrumental requirements for observing and estimating the brightnesses of astrovariables.

The first thing to say is, if you already own a telescope please do press that one into service for your observing programme. It is true that some telescopes will be better than others for this work but few telescopes will prove useless. The same is true of binoculars. In fact, there are more than enough astrovariables bright enough for you to study using humble binoculars to keep you busy for the rest of your life! Most people, though, will wish to use a telescope and so most of this chapter is devoted to matters telescopic.

2.1 What type of telescope is best?

Refracting telescopes, reflecting telescopes, and catadioptric telescopes can all be used for observing astrovariables. Whatever the telescope, it is always desirable to see a field of view of at least 1° when the lowest-power eyepiece is plugged in. This requirement favours a low focal ratio, preferably less than $f/6$, especially for large aperture telescopes and/or the cheaper types of eyepiece.

To see why, let us take an example. The image scale, expressed in arcseconds per millimetre, at the principal focus (no additional optics) of a telescope is given by:

$$\text{image scale} = \frac{206265}{f},$$

where f is the focal length of the telescope in millimetres. A 150 mm $f/4$ telescope has a focal length of 600 mm and an image scale of 344 arcseconds per millimetre. A 150 mm $f/8$ telescope has twice the focal length and an image scale of 172 arcseconds per millimetre.

2.1 What type of telescope is best?

How much of this image is presented to the eye of the observer depends on the size of the field stop aperture of the eyepiece. Let us say that we are using one of the standard $1\frac{1}{4}$ inch (31.7 mm) barrel-diameter eyepieces. Further suppose that it has a field stop aperture of 28 mm (it obviously has to be at least a little less than the size of the barrel). What field of view will we see when we look through it? Plugged into the 150 mm $f/4$ telescope we will see a field of view amounting to 28 × 344 arcseconds, or 9632 arcseconds. Divide this figure by 3600 if you prefer the answer in degrees. This is a field of view of $2°.68$, or about 5 times the diameter of the full Moon.

With the 150 mm $f/8$ telescope the diameter of the field of view we can see is halved (though, of course, the image we see will be magnified twice as much). Our field of view now only amounts to about $2\frac{1}{2}$ times the diameter of the full Moon. That is one quarter of the area it was previously.

A field stop aperture of 28 mm is just about as large as we can get in any eyepiece assembly that has to fit into a standard $1\frac{1}{4}$ inch barrel; so you see the advantage of using a telescope of low focal ratio when we wish to image large fields of view – and the importance of choosing the correct eyepiece. There is more about eyepieces in Section 2.3

Nowadays it is common for refracting telescopes to have fairly low focal ratios ($f/6$–$f/9$), though if they have proper apochromatic object glasses (almost mandatory for low focal ratios in all but the smallest apertures) they are very expensive. By far the cheapest option is the Newtonian reflecting telescope. It can have a focal ratio of below $f/4.5$ and still be relatively cheap, although the very lowest focal ratios are hard to manufacture while still maintaining quality. An $f/5$–$f/7$ reflector of 150–250 mm aperture will normally suit both your needs and your pocket best. It is also true that the larger the aperture the lower is the desired focal ratio in order that the instrument does not become too unwieldy and that its focal length does not preclude a fairly wide field of view when a suitable eyepiece is plugged into it.

Another advantage of reflecting optics is that they do not produce any chromatic aberration. There is always some noticeable secondary spectrum (uncorrected colour) in all but the most expensive of the refractors and this can cause problems when you are trying to compare the brightnesses of stars of differing colour. This is because only a small range of wavelengths (colours) of light are brought to a sharp focus at once. In a normally focused image you will see white stars most sharply defined. Reddish stars along with the bluest stars (and many variable stars are reddish) will then appear very slightly out of focus. True, this will be hard to see but the effect of the slightly expanded disks of the strongly coloured stars will be enough to make them *seem* slightly less bright than they would otherwise appear. The eyepiece which, of course, uses lenses may also give significant trouble in this respect, especially if an inappropriate eyepiece design is chosen – again, see Section 2.3 for more on this.

All practical optics are an exercise in compromise. There is no perfect telescope. However, Newtonian reflectors, as well as being cheap, do have various characteristics that lend themselves very well to most activities of the observational amateur astronomer. I especially recommend this type of telescope for astrovariable observing.

If I was looking for some aspect to criticise, I would say it would have to be that the usual paraboloidal primary mirror is afflicted with a particular optical aberration called *coma*. If the mirror is a good one, it will produce good images close to the centre of the field of view. Stars will look as they should do, intense points of light, near the centre of the field of view. Stars away from the centre will be stretched out into small comet-like shapes (with the tails fanning outwards). The effect of this aberration increases with distance from the centre of the field of view. This is another source of trouble for variable star work. Stars across the field of view will be apparently dimmed by their expanded images. It is the fact that this apparent dimming depends on the position of the star in the field of view that is the real hazard when comparing star brightnesses.

The field of view of acceptable quality, expressed in degrees, that a paraboloidal primary mirror can produce is inversely proportional to its diameter. The following relation can be used as a guide to the maximum diameter, L, of field of acceptable definition for normal visual observing at low magnifications:

$$L = \frac{400}{D},$$

where D is the diameter (or clear aperture) of the primary mirror in millimetres. Thus our 200 mm Newtonian reflector can potentially image a field of view of $2°$ diameter with the star images seen at the edge of the field of view being only slightly distorted. However, for the purposes of making variable star estimates this limit ought really to be reduced by another 15–20 per cent. Refractors are also troubled by outfield aberrations, though usually less so than is the case for the Newtonian reflector.

Fortunately, the limited coma-free fields of Newtonian reflectors of focal ratios $f/5$ and lower can be expanded by means of a correcting lens plugged in just before the eyepiece (in the same manner a Barlow lens is normally used). An example is TeleVue's 'Paracorr' lens, though this unit does produce a $1.15\times$ increase in the magnification, and so reduces the field of view an eyepiece gives to $0.87\times$ the value without the corrector.

On the subject of 'plug-ins', I should mention that a number of companies market telecompressor lenses. These are especially helpful if your telescope is one of the highly popular $f/10$ Schmidt–Cassegrain instruments. They will effectively reduce the focal ratio of a telescope, from say $f/10$ to $f/6.3$ but usually at the expense of some *vignetting* (darkening of the outer parts of the field of view) and an increase in the outfield aberrations. Both of these effects are highly undesirable for our variable star work – so you don't get the increase in the *usable* field of view you might expect.

On the other hand, most catadioptric telescopes and some refractors have markedly curved focal planes, leading to only one zone in the visible field being in perfect focus at once. A manufacturer's focal reducer, designed specifically for that range of instruments, usually flattens the focal plane along with its other function. If you do go in for one of these, be especially sure that it is one intended for visual use. Those giving greater reductions (say reducing $f/10$ to $f/3.3$) will only fully illuminate a small CCD chip with an acceptable image quality.

Of real note are Meade's current series of Schmidt–Newtonian telescopes: a 152 mm $f/5$; a 203 mm $f/4$; and a 254 mm $f/4$. These are basically low-focal-ratio Newtonian's with spherical primary mirrors rather than the usual paraboloids. The corrector plate at the 'sky-end' of the optical tube assembly deals with the spherical aberration that would otherwise afflict the images. Even the 254 mm telescope costs less than $1K in the USA (2003 prices). The advertisements state that the coma afflicting them is half the extent of that for the equivalent Newtonians. As such they should make excellent wide-field telescopes but do bear in mind that their low focal ratios demand expensive eyepieces if they are to provide quality imaging.

When it comes to the mechanics of mounting the telescope, a simple altazimuth mounting, such as a Dobsonian, will suffice for almost any tasks involving visual observation, especially so when using fairly low magnifications. They are aperture-for-aperture the cheapest telescopes you can buy. At the time I am writing these words, the United Kingdom's most productive variable star observer employing visual means is Gary Poyner. He uses his 405 mm Dobsonian-mounted Newtonian reflector to make a staggering 12 000 star brightness determinations per year!

While I have extolled the virtues of the low-focal-ratio Newtonian reflector for observing astrovariables, I must also make it clear that almost any size and type of telescope can produce worthwhile results. The most popular of newly purchased telescopes today are the lines of Schmidt–Cassegrain and Maksutov telescopes marketed by Celestron and by Meade. They are fairly expensive, compared with other options, but they are wonderfully compact and are even fairly portable below about 250 mm aperture. Some people consider a 250 mm or 280 mm Schmidt–Cassegrain telescope to be portable. I do not – at least not easily – but I agree that the 200 mm version certainly is.

Despite all the foregoing, there is one important truism you should remember: the most useful telescope to you is the one you already own!

2.2 What size of telescope is best?

You might think the answer to this is 'the bigger the better'. Actually, for making visual brightness estimates this is not always the case. It could even be that you will need to stop down your large telescope to a smaller effective aperture for all but the faintest astrovariables!

A star produces a *point image*. The perceived brightness of this point image is determined mainly by the aperture of the telescope. Focal ratio has no effect at all provided the optics, including the eyepieces, are all of excellent quality.

Table 2.1 *Telescopic limiting magnitudes*

Telescope aperture (inches/mm)	Limiting magnitude – high magnification	Limiting magnitude – 5 mm exit pupil
1.4/35	11.3	9.9
2/51	12.0	10.6
2.4/60	12.3	11.0
3/76	12.8	11.4
4/102	13.3	12.0
5/127	13.8	12.5
6/152	14.1	12.9
8/203	14.7	13.4
10/254	15.1	13.9
12/305	15.4	14.3
14/356	15.7	14.6
16/406	16.0	14.9
18/457	16.2	15.1
20/508	16.4	15.3

Most practical observers agree that if you observe from a good, dark site and experience a night sky of really excellent transparency then you will see stars of the 6th magnitude. Indeed, if your visual acuity is even slightly better than the average then you ought to be able to see stars as faint as $6^m.5$ on such a night from such a site. Use a telescope with an appropriate magnification and you will see stars which are much fainter than that. How faint? Well, that is a matter of some conjecture. There are a number of old predictive formulae. These give widely different results.

However, Bradley E. Schaefer of the NASA-Goddard Space Flight Center conducted a wide practical survey of contemporary telescope users. He published his results in the 1989 November issue of *Sky & Telescope* magazine. He found that provided the magnification is high enough (perhaps ×200 for a 150 mm aperture telescope, increasing to perhaps ×300 for a 400 mm aperture) one can do rather better than most of the older formulae predict. I have previously proposed (in various of my books and a short paper in the *Journal of the British Astronomical Association* – Volume 107 No. 2) a formula which fits the results of Schaefer's survey very well:

practical limit m_v (high magnification) $= 4.5 + 4.4 \log D$

where D is the telescope aperture, in millimetres, and m_v (high magnification) is the faintest stellar limiting magnitude attainable with the telescope. Table 2.1 shows the stellar limiting magnitudes achievable with a selection of telescope apertures, as predicted by this formula.

The reason for the magnification being important is that the star image is seen against the image that the telescope produces of the sky background – *and this background is not perfectly black.*

2.2 What size of telescope is best?

The sky background behaves like an *extended image*, meaning that the light from it is spread out, as opposed to being concentrated into a tiny point. The telescope gathers an amount of light determined by its aperture and presents this in the image that we see when we look through the eyepiece. Use a given magnification and the star will be seen as a point of light. Even from a site not bedevilled by light pollution, the surrounding sky will appear as a grey wash of light covering the field of view. Double that magnification, though, and the light in a given patch of sky background will be stretched to four times its previous area and so will appear only one quarter as bright as before. In general, the brightness of the sky background is inversely proportional to the square of the magnification.

Using more magnification effectively darkens the sky background while the star image is little affected (until too much magnification is applied, when its image ceases to be point-like). The better *contrast* between the star and its background at the higher magnification makes it *seem* brighter. Image contrast is a crucially important factor when trying to see things. Hence, up to a point, increasing the magnification used on a given telescope makes the stars we see through it appear more distinct. It also enables one to detect stars that may be lost in the background sky glow at lower magnifications. Thought of in this way, Schaefer's results hardly seem surprising.

We must remember that the foregoing formula predicts the magnitude of the faintest star observable with a given aperture telescope – and a high magnification is needed to realise that faintest limit. Often we will need a field of view large enough to encompasses the astrovariable along with several comparison stars. In order to get that field of view we will more usually need to use a much lower magnification than that which will allow us to see the very faintest stars.

However, we cannot make the magnification too low. Just how low depends partly upon the aperture of the telescope. The eyepiece of a telescope produces a disk of light, usually known as the *exit pupil*, or the *Ramsden disk*, or sometimes as the *eye ring*, which is, in effect, an image of the telescope objective formed by the eyepiece. The exit pupil is situated just a little distance, known as the *eye relief*, from the eye lens of the eyepiece (see Figure 2.1). All of the light from distant objects which is collected by the objective of the telescope passes through the exit pupil. The observer automatically steers his/her eye pupil to coincide with the exit pupil in order to receive the maximum amount of the light collected by the telescope.

The size of the exit pupil is inversely proportional to the magnification the eyepiece produces with the telescope. In fact, we can calculate the size of the exit pupil from the following relation:

$$\text{diameter of exit pupil} = \frac{\text{aperture}}{\text{magnification}}$$

The diameter of the exit pupil, and the aperture of the telescope have to be given in the same units. As an example, if a 250 mm reflecting telescope has an eyepiece

Variables in vision

Figure 2.1 The bright image of the primary mirror of the author's telescope can be clearly seen in the eyelens of the eyepiece. This image is the exit pupil, the best position for the observer's eye in order to see all the celestial light collected by the mirror. The image is actually centred on the optical axis but is formed a short distance in front of the eyelens, which is why is appears displaced to the upper-left in this oblique view.

plugged into it to produce a magnification of ×50, then the exit pupil produced has a diameter of 5 mm. The eye pupil of a teenager can expand to about 7 mm or 8 mm when fully dark adapted. As one ages this figure decreases. The eye pupil of an average 30 year old person will expand to around 6.5 mm and this figure decreases further by about 0.5 mm per decade. Thus an average 60 year old has an eye pupil that will open to only about 5 mm at best.

If the pupil of the observer's eye is smaller than the size of the exit pupil, then not all of the light collected by the telescope can enter his/her eye in one go. Putting it the other way round: the observer must choose a magnification in

2.2 What size of telescope is best?

order to make sure that the exit pupil is no bigger than the his/her eye pupil if he/she is not to waste some of the light grasp of the telescope.

Consequently, there is a minimum magnification you should use on a telescope. Use a lower magnification than that and you will waste some of the precious celestial light collected by it. If you are a 60 year old and are using a 250 mm aperture telescope, then you will be wasting some of its light-grasp if you use an eyepiece with it that gives a magnification of less than ×50.

What does 'wasting light' mean in practice? Obviously, anything in the image that is point-like, for instance the image of a star, will seem less bright than it should. Indeed, the faintest point images might be then rendered too dim to see at all. It is just as if the telescope has had its aperture reduced, or it has been *stopped down*, as we say.

The effect on an extended image is a little different but I will not discuss this here as we are only concerned with estimating the brightnesses of stars and star-like objects.

There is another reason why an overlarge exit pupil is not desirable for any instrument with an obstruction in the light path (which happens to be the case for almost all forms of telescope). Take another look at Figure 2.1. The exit pupil has silhouetted within it the obstruction due to the secondary mirror of the telescope (in this case a 0.46 m Newtonian reflector, with a secondary obstruction spanning a quarter of the diameter of the primary mirror). Normally, the observer is not aware of the secondary's silhouette but when the exit pupil is large enough, the silhouette can then be a substantial portion of the size of the observer's eye pupil. Then it does make itself felt by producing an unpleasant shadowing effect. So, if you use a Newtonian reflector you really are limited to magnifications not much lower than the aperture of the telescope divided by 7 or 8. Of course, using a magnification low enough to produce an overlarge exit pupil is always wasteful of the precious celestial light but one can still do so with the unobstructed refractor without incurring any extra penalties.

Having considered the lowest magnification we can use on a given telescope, what limiting stellar magnitude can we expect from it when we use a low magnification?

Once again Bradley E. Schaefer's survey can help us. I have been able to derive another predictive formula from his published results which relates the limiting magnitude to telescope aperture when the magnification is such as to produce an exit pupil of 5 mm diameter (and hence a magnification numerically equal to the aperture of the telescope in millimetres divided by 5):

practical limit m_v (5 mm exit pupil) $= 2.6 + 4.7 \log D$

Table 2.1 includes a column of faintest magnitudes predicted from my formula when such a low magnification is used.

Why a 5 mm exit pupil? In my view a magnification that produces an exit pupil of this size is about right for low-power viewing. I say this in the face of those who insist that a 7 mm exit pupil is best. With such a low magnification the sky background will look a light grey on all but the most crystal-clear nights and the visibility of both stellar and faint nebulous objects will greatly suffer as a result. In addition, there is the concern about wasting light discussed earlier.

Of course, any formulae such as these can only ever be a rough guide because there are many factors which will affect the final result. The predictions from the 'high-magnification' formula are especially uncertain for large apertures as these are most affected by adverse seeing conditions. Unsteady seeing will cause the seeing disks of the stars to blur enough to render the faint ones invisible. The predictions of the '5 mm exit pupil' formula are rendered particularly uncertain when any combination of moonlight, 'light pollution', and haze causes the sky background to brighten. Despite these qualifications, based as they are on a real practical survey of modern-day telescope users, these formulae are a much better guide to what you can expect from your telescope than other versions you will find commonly in print.

So, we can use the formula to decide the *minimum* aperture of telescope needed to observe a star of a certain brightness. However, if we wish to make an accurate visual estimate of its brightness there is also a *maximum* aperture we should consider using. The reason for this lies with a characteristic of our eyes. Our eyes have a hard enough time in comparing brightnesses of star-like images when they are not too bright. This difficult task becomes even tougher when the star-like images are bright.

It turns out that for us to have the best chance of achieving the desired $\pm 0^m.1$ accuracy the star image should ideally be 2–3 magnitudes brighter than the limit of visibility. So, if you wish to accurately determine the brightness of a star whose magnitude you know to be somewhere about $11^m.5$, then a telescope aperture that allows you to only just discern stars of magnitude 14^m would be a good choice. On a night of good transparency you could achieve the desired brightness with an aperture of about 150 mm if you select the optimum magnification to squeeze maximum performance from the telescope (about ×200).

This might be OK if the comparison stars are all very close to the astrovariable. If you have to use a lower magnification in order to gain sufficient field of view then the faintness limit will be raised and you will need a larger aperture to achieve a star image of ideal brightness. For instance, if you use a magnification that produces an exit pupil of about 5 mm, then a telescope aperture of something around 250 mm would be best to visually estimate the brightness of our hypothetical $11^m.5$ astrovariable.

If you own a telescope of, say, 200 mm aperture there is nothing you can do to it in order to turn it into one of larger aperture. However, you can always make a diaphragm to stop it down to a smaller aperture for those occasions when you need to.

You can make diaphragms from cardboard with a range of apertures to suit various astrovariables (and perhaps even varying observing conditions). They can be placed over the sky end of the telescope with hooks to secure them in place. In the case of a reflecting telescope the diaphragms can be also be placed close to the primary mirror if the telescope has a skeleton tube or an access hatch close to the mirror.

Do not obsess about having a large number of diaphragms to suit every occasion, though. You can make accurate brightness estimates for astrovariables that are more than a magnitude either brighter or dimmer than the ideal. So, even if you are lucky enough to have a telescope of about half a metre aperture, four diaphragms will be sufficient. Each one could halve the aperture. Their sizes might then be: 250 mm, 125 mm, 63 mm, 31 mm. Their effect on the limiting magnitudes will be sequentially circa $1^m.3$ in actual practice, even though simple theory predicts a larger jump (consult Table 2.1 for the practical result).

Where the telescope has no secondary obstruction, as is the case with a refractor, the hole should be made on-axis (at the centre of the disk). For telescopes, such as catadioptrics and most reflectors, which do have a central obstruction the hole can still be central until the central obstruction would become about half the diameter of the clear aperture. Then the hole in the cardboard disk should be made off-axis, so that it misses the central obstruction. In that case, it would also be good to position the hole so that it avoids the secondary support vanes (in the case of a Cassegrain or Newtonian reflector). This is simply to avoid diffraction bars extending from the star images – a minor point but I think it is always desirable to get the best quality images from any equipment.

Unless your telescope is one where you can image a couple of degrees or more of sky, it should be equipped with some form of finder telescope. The highest optical quality is not needed but it should have as large an aperture as possible and it should be fitted with an eyepiece giving a real field of at least $2°.5$. The magnification should ideally be equal to its clear aperture in millimetres divided by 5, so producing a 5 mm exit pupil. Crosswires are a help only if they are illuminated. A red LED (light-emitting diode) might do as the source of illumination, if you decide to install illumination in a cheap finderscope yourself (only to be attempted if you are sure your are up to this task – delicate optical devices are *very* easily ruined!).

After all the foregoing I still have not squarely answered the original question: what size of telescope is best? The answer to that depends on how faint the astrovariables are that you wish to study. Table 2.1 will help you decide what aperture you need for given astrovariable brightnesses. It is worth emphasising, though, that even a modest 50 mm aperture will allow you to study thousands of objects, so a pair of binoculars or a small refractor is ample enough to serve you for a lifetime's observing!

2.3 Eyepieces and fields of view

For our work we will often need a field of view of a degree or so. This is to enable us to first find our chosen astrovariable, and then to see it along with some chosen *comparison stars*. Comparison stars are ones of known magnitude which we can compare with the astrovariable in order to estimate its brightness.

To get a large field of view we will need to use a lowish magnification. What type and size of eyepiece should we use and what size of field of view can we expect from our chosen eyepiece? The following notes address these questions.

It is worth taking some care when one is choosing a particular telescope to purchase, given the almost bewildering array available nowadays. This is even more true when choosing the eyepieces to use with it. The modern generation of eyepieces can swallow up a sizeable chunk of the budget for the whole installation. Few appreciate until it is too late that choosing the correct eyepiece(s) for the observational task is just as important as choosing the most suitable telescope.

One important numerical relationship that is worth giving here is:

$$\text{real field} = \frac{\text{apparent field}}{\text{magnification}}$$

In using this equation, the values of real and apparent field must be in the same units. Usually they are both expressed in degrees.

For most practical purposes the *apparent field* of the eyepiece can be defined as the angle through which the observer's line of sight must swivel in order to see from one extreme edge of the field of view to the other. This quantity is a characteristic determined by the design of the eyepiece.

As you will be aware, apparent fields of view of more than 55° come from complex designs of five- to eight-element eyepieces – such as the TeleVue company's 'Radian' (60°), 'Panoptic' (68°), and 'Nagler' (82°), and Meade's 'Super Wide Angle' (67°), and 'Ultra Wide Angle' (84°). These cost hundreds of dollars each, so you may well have to set your sights lower. 'Plössyl' eyepieces are very easily available from a number of manufacturers (mostly in four-element designs, sometimes five-element) and they usually have apparent fields of view in the range of 50–55° and cost circa $80 at current (2003) prices.

The *real field* factor in the equation is the angular extent of the sky that the observer actually sees when he/she uses a particular eyepiece with a particular telescope. The diameter of the full Moon subtends just over $\frac{1}{2}°$, for instance, so you will need a real field of at least this size in order to see all the full Moon in one go. As an example, if an eyepiece of 60° apparent field of view gives you a magnification of ×40 when it is plugged into your telescope you will see a real field of 1°.5 (almost three Moon diameters).

One thing I should warn you about is that you may well get a slightly smaller real field in practice than you might expect using the equation with the manufacturer's stated value of apparent field for the eyepiece. Almost all eyepieces suffer to a small degree from pincushion distortion. That is, the magnification of the image increases a little away from the centre of the field of view. This aberration is usually worst in eyepieces of large apparent field. For instance, a Nagler eyepiece, with its stated '82°' apparent field, will behave like an eyepiece with an apparent field closer to 78°, as far as the equation predicting the real field is concerned.

The effects of chromatic aberration and the various seidal aberrations (spherical aberration, coma, astigmatism, curvature of field, and distortion) will make themselves felt if you use an eyepiece of too simple a design on a low focal ratio telescope. 'Achromatic Ramsdens' and 'Kellners' are just about acceptable for low and moderate power views with telescopes of focal ratio not much lower than $f/6$. I say this in the face of the fact that manufacturers often supply Kellner eyepieces with $f/4.5$ Dobsonian telescopes! They have fields of view of around 40°. Their one virtue is that they are cheap.

A Plössyl eyepiece is only just about good enough for very-low-power views (the eyepiece having a focal length of not much less than 25 mm) on a telescope of focal ratio around $f/4$. Even with low powers, star images in the outer third of the field of view will appear noticeably aberrated and the portion of the field of view suitable for the astrovariable and its comparison stars spans just under the remaining two thirds (roughly the centremost 32° of the typically 52° field of view). The situation is much better at $f/5$, the Plössyl then performing well at low and moderate powers with only a little image degradation apparent near the edge of the field of view. At $f/6$ it will deliver good images to quite near to the edge of the field of view over the full range of magnifications.

The best Orthoscopic eyepieces can out-do the performance of the Plössyl type, though the apparent field of view is a little smaller, being in the range of 40–45°.

Most of the more complex designs will work well with low-focal-ratio telescopes, though the quality of the images near the edge of the field of view will suffer at the very lowest ratios even with these eyepieces. The expensive, but phenomenal, Nagler eyepieces are the champions at focal ratios as low as $f/4$, though Radians are also excellent at low focal ratios. Radians are especially notable for the high contrast views they deliver.

When you are selecting equipment for your observing, I recommend that you consider the eyepieces in combination with the telescope, rather than treating them as an afterthought. In particular, consider the cost of the eyepieces you will need when selecting the focal ratio of the telescope. You will be able to achieve the biggest field of view at the smallest cost if you balance the focal ratio against the eyepiece type needed.

2.4 Vignetting

Unless there is something wrong with the design, construction, or collimation of your telescope, the in-focus star images at the centre of the field of view ought to be created from the intersection of all of the rays of light from the primary mirror/object glass. This is the same as saying that the centre of the field of view is fully illuminated by the telescope's main light collecting component.

For our work of comparing the brightnesses of stars simultaneously situated at different positions across the field of view, we also need the outfield stars to be fully illuminated by light from the primary mirror/object glass. For instance, if one or more obstructions in the light path cause a given outfield star's light to be formed from the rays of light from only 90 per cent of the area of the primary mirror it will appear only 90 per cent of its true brightness. We say that the obstruction *vignettes* some of the rays. In this hypothetical case vignetting would result in an artificial dimming of the star by 10 per cent of its true brightness. In magnitude terms this is equivalent to a dimming of the star by $0^m.1$. Meanwhile a star at the centre of the field of view will be seen at its full brightness.

A dimming of $0^m.1$ is just detectable by the human eye and we are hoping to estimate star brightnesses to $\pm 0^m.1$. Consequently it is important that all parts of the field of view that we intend utilising for our star comparisons suffer vignetting of rather less than 10 per cent.

How does your telescope rate in this respect? There are two ways to go about answering this question. One is by a simple practical trial to judge the presence of vignetting qualitatively. The other is by making some simple measurements of the parts of your telescope and performing some calculations. The results of those calculations will also enable you to make an informed judgement about making any modifications to the affected parts of your telescope, and if so by how much.

We start with the practical trial. This assumes that your telescope is properly collimated. Refer to Section 3.2 for details about collimating your telescope.

First determine the position of the focal plane of the telescope. A night-time test using the Moon or a bright planet or star is probably the safest way of doing this. Put a small piece of thin paper (or other translucent material) across the mouth of the focuser drawtube and adjust until the image of the Moon, or whatever bright object, is in sharp focus. Either mark the drawtube, or measure the distance of the top of the drawtube from the base plate in order for you to place the mouth of the drawtube in the same position during the daytime. Alternatively you could continue the test at night but you will need to illuminate a wall or some form of screen and point the telescope at it.

Adjust the focuser so that the mouth of the drawtube coincides with the telescope's focal plane (the plane in which you found the image of the Moon, or other object, to be in focus). Put your eye to the mouth of the drawtube. With your eye pupil positioned near the middle of the drawtube opening you should get a clear view of the primary mirror/object glass of your telescope. Do you? If so, you can be sure that the centre of the field of view is fully illuminated.

Now move your head about slightly so that your eye can search across the end of the drawtube. Do you still get a clear view of the primary optic? Or, perhaps when your eye pupil is placed near to the side of the opening, do you see a little of the primary optic obscured by something? What is that something? Is it something you can change?

How big is the area in the focal plane which suffers only very little vignetting? To find out this make up a small cardboard disk with a central hole cut in it. Tape the disk to the mouth of the drawtube. Make the size of the hole in it

2.4 Vignetting

equal to your best guess at the size of the region from which you see little or no obstruction, based on your rough trial. Try putting your eye to this and search out the hole to see if you see any obstruction.

Make a second cardboard disk with the hole size based on the result from the first one. When you have got the size of the hole correct you will see only a slight obscuration of the primary optic when your eye pupil is placed at the extreme edge of it. This hole now defines the field of view of your telescope within which any vignetting can be tolerated for your astrovariable work.

If the field stop aperture of whatever eyepiece you intend using for your astrovariable work is bigger than the size of the hole in the cardboard disk you can be sure that the outer parts of the field of view are going to be affected by vignetting by more than is tolerable for our astrovariable work. I should mention here that in almost all other modes of observing this amount of vignetting is likely to be completely unnoticeable – it is just that we have to be unusually critical in our requirements for our astrovariable work!

Do bear in mind, though, that the outer 10–20 per cent of the field of view is likely to be unusable for our purposes because of eyepiece outfield aberrations anyway (and in some cases optical aberrations, such as coma, from the telescope itself).

In most types of eyepieces (Huygenian and Huygens–Mittenzwey eyepieces being the only common exceptions) the field stop aperture is situated just a little inside the barrel, and so is available for your inspection and measurement. Compare it with the size of the hole in your cardboard disk.

Alternatively you can measure the size of the hole in the cardboard disk and from that determine the size of the (almost) unvignetted field of view of your telescope (using the formula I gave for a telescope's image scale in Section 2.1). This figure can be compared to the sizes of the real field of view of your preferred eyepiece(s), which you can calculate using the formula I give in Section 2.3. You will certainly have to do this if your eyepiece does not have an accessible field stop.

As I previously mentioned, the other way to investigate the presence of vignetting is by measuring some of the components of your telescope. One possible source of vignetting is the secondary mirror present in most types of telescope. See Figure 2.2 for the optical layouts of Newtonian and Cassegrain reflectors and Schmidt–Cassegrain telescopes. Though many thousands of Cassegrain telescopes were made, they are only rarely found in the hands of today's active amateurs. Still, I will include brief details here for those readers who do use them.

Maksutov telescopes are similar to Schmidt–Cassegrains when it comes to investigating for vignetting, as are Schmidt–Newtonians similar to Newtonian telescopes in this respect. However, it is only in Newtonian and Cassegrain telescopes that the secondary mirrors are accessible for you to change if you so desire. If I had paid several thousand dollars for a factory-made, catadioptric telescope that would, for me, certainly be the end of the matter!

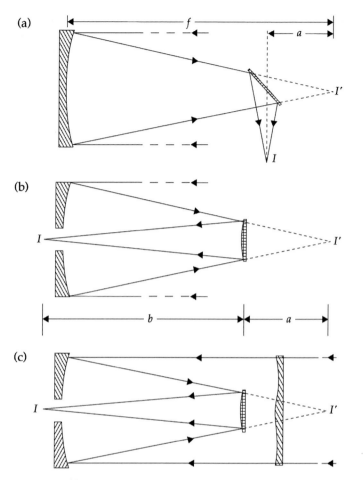

Figure 2.2 (a) The Newtonian reflector. (b) The Cassegrain reflector. (c) The Schmidt–Cassegrain telescope. The quantities shown can be used in the equations given in the text to check for vignetting.

Vignetting by the Newtonian secondary mirror

Referring to Figure 2.2(a), if it were not for the secondary mirror the rays from the primary mirror would form the image at I'. Instead the rays are diverted to an accessible position to form the image at I. The intersection point of the secondary's reflective surface and the optical axis is placed a distance a inside the primary's focus. The final focus at I is this same distance from the intersection point. By measuring the distance of the focal plane from the side wall of the telescope tube (as described earlier – by imaging a bright object on a piece of paper) and measuring the distance of the centre of the secondary mount to the wall of the telescope tube you can find this distance a for yourself. Then knowing the focal length, f, and the diameter, D, of the primary mirror (look at the diagram to see how you can measure up to check on the manufacturer's quoted values if you wish to) you can make the calculations to check for vignetting.

To illuminate fully just the centre of the field of view the secondary mirror has to have a minor-axis diameter of at least aD/f. To illuminate fully the focal plane across a diameter, d, the secondary mirror should have a minor-axis diameter, w, given by:

$$w = d + \frac{(D-d)}{f}$$

All the quantities in the equation, including w, must be in the same units. I recommend using millimetres throughout.

Based on your findings, you can decide if you wish to order a new secondary mirror for your Newtonian reflector. You will, of course, need a new cell to house it and you may even need a new secondary support to go with it. Bear in mind that even with the same secondary support you may have to fit it slightly further up the telescope tube in order to accommodate the larger size of the secondary mirror. I must repeat the same advice I have given before: *only consider taking on any alterations to your telescope if you are really sure that you know what you are doing and that you can safely store the optics while you are carrying out the mechanical work. Telescopes are expensive items and can be VERY easily ruined!*

Vignetting by the Cassegrain secondary mirror

Referring to Figure 2.2(b), you will notice that the secondary mirror is a distance a inside the focus of the primary mirror and is a distance b from the focal plane. Unlike the flat mirror of the Newtonian reflector, a Cassegrain telescope's secondary mirror is convex in form and actually reduces the convergence of the rays arriving from the primary mirror. Hence after reflection they can travel a large distance, b, before they reach a focus at I. If you were to study the geometry of the situation you would see that the *effective focal length* of the telescope is equal to the focal length of the primary mirror multiplied by the ratio of the distances $b:a$.

Thus:

$$\text{effective focal length} = \frac{fb}{a}$$

and

$$\text{effective focal ratio} = \frac{Fb}{a}$$

where f and F are the focal length and the focal ratio of the primary mirror, respectively.

The minimum diameter of the Cassegrain secondary mirror just needed to illuminate fully the centre of the field of view is Da/f, where D is the diameter of the primary mirror. To illuminate fully the field at the focal plane out to a diameter d the diameter of the secondary mirror, w, has to be:

$$w = a\left[\frac{D}{f} + \frac{d(f-a)}{bf}\right].$$

If you have a Cassegrain telescope and decide you need a new secondary mirror for it you will need to order one of the correct focal length from the optical worker who is going to make it for you. You could either contact the original manufacturer of your telescope (who may be the best person to make you a new secondary mirror, after all) or you will have to calculate its focal length, f_s yourself. Use:

$$f_s = \frac{ab}{(a - b)}$$

Failing that, you will have to send the optician the primary mirror so that he/she can make all the measurements himself/herself and craft you a new secondary mirror to suit. This will, in any case, be necessary if there is any doubt about the exact figure on the primary mirror – for instance, are you sure that your telescope is a classical Cassegrain and not a Dall–Kirkham or a Ritchey–Crétien?

Vignetting caused by drawtubes

The major source of vignetting, at least for the outermost parts of the field of view, will normally be the drawtube. If the end of the drawtube furthest from the eyepiece (or the bottom of the focuser mounting if the drawtube doesn't reach this far in when normally focused) has an internal diameter of B and it has a length A measured from the field stop aperture of the eyepiece plugged into it, the diameter d of the fully illuminated field is given by this approximation:

$$d = \frac{B - A}{F}$$

where F is the effective focal ratio of the telescope. Changing a telescope's 1¼ inch (31.7 mm) standard focuser for a 2 inch model (which can always be fitted with a 1¼ inch adapter) will often solve any problems here.

2.5 Binoculars

A good pair of binoculars can provide you with enough variable stars to keep you busy for the rest of your life. You will only need a telescope if you are determined to go after the fainter ones. Of course, if you can afford the expense you can have the luxury of a telescope *and* a pair of binoculars. However, do beware buying the cheapest models of either. I especially urge you to go for quality above size every time.

Many authorities recommend 7 × 50 binoculars (magnification factor 7, 50 mm diameter object glasses) as the best size you can choose for general astronomical usage. However, I disagree. This aperture to magnification ratio produces an exit pupil of 7.14 mm diameter at each of the eyepieces. This is OK if you are in your teens but is wasteful of light if you are much older than that. For example, if your eye pupils will only expand to 5 mm when fully dark adapted,

2.5 Binoculars

then you might just as well have chosen the equally common 7×35 binoculars. Your views through either will be indistinguishable and the 7×35 binoculars will be much cheaper, as well as more compact and lighter to carry and handle. Remember, also, that the sky background will appear too light at the very lowest magnifications, affecting performance. The binoculars I recommend as best are the 10×50 size. Their 5 mm exit pupils will suit the eyes of people aged from 6 to 60.

You can use the figures in Table 2.1 (in the '5 mm exit pupil' column) to estimate the likely limiting magnitude achievable with binoculars. A good quality pair of 7×35 binoculars ought to show you stars of brightness about $9^m.9$ on a good night and a good pair of 10×50 binoculars ought to allow you to detect stars down to around $10^m.6$.

What about using binoculars with higher magnifications, though, and perhaps binoculars with bigger objective lenses – would this be a good thing to do? Not necessarily.

One advantage of binoculars over telescopes is their inherent wide field of view. Typical midpriced binoculars will have eyepieces fitted that have apparent field diameters of about $50°$, or maybe just a little larger. A magnification of $\times 10$ will result in a real field diameter of $5°$. That is, for instance, just enough to squeeze all the main stars of the Hyades star cluster into the field of view in one go. Binoculars in the same price range giving higher magnifications necessarily have smaller fields of view. The price of binoculars fitted with eyepieces of much wider apparent fields really starts to rocket upwards. So, not only would you have to spend more money to buy more magnification, you would be sacrificing some of the main advantage of binoculars over telescopes: field size.

There is another disadvantage to binoculars of higher magnification. You will find that holding them steady enough is not at all easy. A few moments of trying to fix your eye on the subject while it jitters and swims erratically about in your vision will make you feel queasy. Not only that, your eye and brain will have precious little opportunity actually to study the subject and your variable star estimates would have a likely uncertainty of nearer 1 magnitude than 0.1! You would have been better choosing the lower-power binoculars instead, and saving some money into the bargain.

Large binoculars are also *heavy* binoculars. How long will you be able to hold a hefty pair of binoculars in position before the ache in your arms and shoulders, soon spreading to your neck, gets too much to ignore? Not very long.

Even with 10×50 binoculars, you will soon see the advantage of resting your elbows on a suitable platform when using them for long periods. The top ledge of a fence, or the platform of a bird-table or anything else you can press into service will increase your comfort – and hence the detail you will be able to appreciate and observe – when using such a rest to ease your muscle load.

If you are to achieve accuracy in your magnitude estimates your binoculars really ought to be provided with some sort of stand. This does open up the opportunity to use binoculars of greater aperture and magnification if so desired. A number of different mountings are commercially available. Some work on the same general principle as the table-lamps of the famous 'Anglepoise' tradename. Some have counterweights, others have springs to do the same job. In some cases you can attach these units to the arms or frames of deck-chairs, making for very comfortable viewing. I recommend you scan the adverts if one of these appeals to you.

Big mounted binoculars are extremely expensive but you do get a quality instrument for your money. For instance, Fujinon's 25×150 binoculars come on a sturdy tripod and tilt-and-pan head. They also have extra prisms incorporated in the optical train in order to have the eyepieces angled upwards by $45°$. There are some others angled upwards to $90°$. This feature affords a great relief on the neck muscles when viewing any object at high altitude (*altitude* here referring to the angular distance of the object above the horizon).

Binoculars, large and small, can be found second hand but please do examine them closely (particularly looking for scratches in the lenses and checking that the focusing action is smooth and slop-free) and thoroughly check them out in use before you part with any money.

You should particularly carefully check the alignment of each half of the binoculars. With the binoculars looking at some far away object, look through the left half with your right eye closed. Then close your left eye and open your right eye. Is there any change in the position of the image within the eyepiece field? If so a prism might have been knocked out of line. I have managed to realign a pair of faulty binoculars myself but it is a tricky thing to do without the right equipment. In any case, and sorry to say it yet again, *please do not even consider taking on the dismantling of any piece of expensive optical equipment unless you are sure that you are up to the task*. You have got to know what you are doing and you must work with scrupulous cleanliness otherwise the performance of the binoculars, or other equipment, may well be significantly impaired.

When buying the same basic advice goes for any optical device, large or small, that you intend to purchase: buyer beware!

Image stabilized binoculars are of particular interest because they give a very steady view at higher magnifications without the need for a tripod. The effect of the image stabilization technology when hand-holding higher-power binoculars is impressive. When switched off, objects jitter around in the field of view just as they would in conventional binoculars. Press the stabilization button, however, and the view becomes almost as good as it would be if the binoculars were solidly mounted. The image still moves but the motion is now silky smooth. The detail-destroying jittering is vanquished. On a practical trial I found a definite improvement in visual limiting magnitudes and perceived details over non-stabilized binoculars of the same aperture. The downside is that image stabilized binoculars are very expensive.

Chapter 3
Astrovariables reckoned

This chapter details how to make magnitude estimates of astrovariables using the human eye and brain as the detector and measuring device. This usually will be with the aid of additional optical equipment such as binoculars or a telescope, though observations made with the eyes alone are also possible for a limited number of stars.

You might think this to be a very 'low-tech' approach in these times of computerised 'GO-TO' telescopes and CCD astrocameras but the honest truth is that most observers of variable stars still work in this 'old-fashioned' way.

Yes, of course changes are afoot. In the years to come ever larger numbers of amateur practitioners will be using CCDs and their associated computer software instead of their eyes and brains. For the next few years, though, the majority of your observing colleagues will still be using purely visual methods to monitor variable stars. As long as this situation lasts you can still make a truly valuable contribution by doing the same.

3.1 Preparations

In order for you to experience the greater pleasure while you are observing and to achieve the most from each observing session, it is as well to spend a small amount of time and effort making some plans and some preparations. Some of these will be made well in advance, while others will be made just prior to going out to your telescope.

One piece of preparation that will only have to be made once is the construction of some sort of illuminated clip-board to carry notepaper for use at the telescope. A small piece of hardboard, or a stout dinner mat, a switch, a torch bulb and holder, a small square cardboard or metal box (to mount the bulb in its holder at the head of the board, the box being open and facing down the board), a switch, some wire, a battery (perhaps mounted on the board by means of a

Terry-clip) or wires to connect to an external LT supply, and terminal connections (or soldered joints) can easily be fashioned into something appropriate. A small rheostat would be a good addition so that the brightness of the illumination can be varied. Incorporating a shield to cut off the direct light from the bulb entering your eye would be good. It is also advisable to have the bulb painted red or covered with a red filter as red light will disturb your dark adaption less than unfiltered light. Keep the construction of the board simple, though, and above all keep it lightweight.

Another useful provision is a clock (or watch) set to Universal Time (UT). This is especially so if you are observing from a location whose Zone Time is several hours removed from UT. The various local provisions for Daylight Saving Time, Summer Time, etc. are also nuisances. Failing a UT clock, I would advise recording your observations on your notepad in the current Zone Time. Don't burden yourself with having to convert the times while you are observing. Do the conversions afterwards. You can see the convenience afforded by a clock or watch set to UT. In any event, the observing group(s) you submit your observations to will want the times recorded in UT. It may be that you will want to (or be required to) submit the date and time of the observation in terms of its Julian Date. If so, this is something that will have to be derived after the observation and that relies on you knowing the UT, rather than your Zone Time.

Eventually you will develop a portfolio of the astrovariables you wish to monitor. I would advise you to make notes about each of the objects that interests you. Any information you can find about their maximum and minimum brightnesses (or predicted, or likely, brightnesses) is especially useful. You can relate this to the equipment you have at your disposal. To take the most obvious example, it will be no good you deciding to observe a variable star which can never become bright enough for you to see it in your telescope!

Do seek advice from your observing group co-ordinator as to which objects are suitable for you to study and which are apposite to the observing programme the group is engaged in.

I will not labour the point about creating the portfolio as you will develop this to suit your needs and interests as you go on. Keep a hard-copy of it because computer memories and floppy disks do become corrupted. You might find that an alphabetical entry of one page per object will suit you best. It is from this (growing) portfolio that you can select which objects you are to observe shortly before you go out to your telescope.

Let your portfolio grow slowly and naturally. When observing, it will take you anything from a few minutes to about an hour to locate and verify (making sure you really have got the correct astrovariable) each new object and make your first brightness determination of it. With practice the whole process will take you less time but do expect things to go slowly at first. What I am saying is do not arm yourself with the charts of fifty stars on your first night of undertaking this work. Be pleased if you do one observation on your first evening and be ecstatic if you achieve four or five!

3.1 Preparations

It is worth making one crucial point here: *Just one observation carefully and accurately carried out is valuable. Fifty hasty and inaccurate observations are worse than useless as erroneous results can spoil any analysis, especially in the cases where there are few contributing observers.*

You eventually acquire dozens, perhaps more than a hundred, astrovariables in your portfolio; so which ones do you observe this evening? Well, there might be an alert issued by your observing group to monitor one or more of them. That is a good start to the night's list, if you decide to join in with the 'call to arms'. Obviously monitoring the appropriate web-page is an especially effective way for you to keep abreast of all the activity in your group.

For more routine monitoring, one important consideration is having your chosen objects well placed in the sky. That usually means having them high in the sky. With experience you will become familiar with what constellations, and so what right ascensions, will ride high during the course of any particular evening. Failing that you can find out either by calculation based on figures from an ephemeris or by running a planetarium computer program.

Any celestial body will reach *upper culmination* (when it is at its greatest altitude above the horizon) when its *right ascension* is equal to the *Local Sidereal Time*. If you are happy about calculating this (only very rough figures are needed to check whether any celestial body is going to be somewhere near upper culmination) then all well and good. If not, but you would like to know how, then I must refer you to Chapter 13 in my book *Astronomy in Depth* (Springer-Verlag, 2003) as we do not have the space to go into this involved subject here.

The easiest alternative of all is to run a computer planetarium program such as *Starry Night Pro* and from the displayed sky decide what right ascensions (and declinations) can be seen to the best advantage from your site in the course of that evening. I observe frequently enough always to know this without needing to do calculations or use my computer. If you do not at this point in your observing career, I expect you soon will, too.

While you have your computer running, do any printing out of whatever finder charts, etc. you will need for your imminent observing session (you will soon develop a file of print-outs all ready to go, perhaps slotted into the notes in your portfolio).

Do as much as you can as early as you can in order that you can go out and observe as soon as possible when it gets dark, when the clouds clear, and when you are rested and ready for your evening's observing. You want to spend your chosen time observing, not in a manic dash trying to find and assemble everything in the hope that you can fit in an observation or two before the clouds roll in.

Wear appropriate warm clothing to suit the ambient temperature of the evening (especially so if you are planning a long session). An uncomfortable observer is an inaccurate observer. Also, you will soon get fed up with going out at night if each time you endure the agonies and agues of the biting cold. In the coldest weather I wear an extra pair of thick woollen socks, sometimes an extra pair of trousers, a Balaclava in addition to a thick woollen hat, as well as the rest of the heavy clothing you would expect me to wear. I may look silly (though who is to see me in the dark?) but I enjoy feeling as warm as toast as I walk across my grass lawn when it is frozen as hard as concrete!

If you have to observe from a site which is removed from your home, then it is a very good idea to create a check-list of all the things you need to take with you. This might include a Thermos of hot drink and a light snack. Do double-check your check-list before you set off. It would be a real disaster if you were to spend half an hour loading up your car, half an hour driving to your chosen site, ten minutes unpacking your car, fifteen minutes setting up – and then discover you forgot to pack your telescope's eyepieces!

One matter I ought to mention under the heading of preparations is the collimation of your telescope. It is such a necessary but involved subject that I have devoted the following section to it. In it I offer brief notes on how to collimate the main forms of telescope you are likely to encounter. Each is described under its own subheading, so you only really need to concern yourself with the advice given for your particular type of telescope.

3.2 Collimation

We require good, tight star images across the major part of the field of view for our task of comparing star brightnesses with astrovariables. In order to achieve this it is crucial that whatever telescope we are using is in good *collimation*, by which I mean that the optical components are accurately positioned and aligned.

It is true that we can put up with any minor aberrations or shortcomings of image quality that afflict star images in the major part of the field of view equally. The stars would then all be expanded or distorted by the same amount and so apparently all dimmed by the same factor. We certainly, though, cannot tolerate the size and shape of a star's image varying to any significant extent with its position in the field of view, accepting there will always be some degradation near to the edge of the field of view.

Some observers insist that every time you go out to your telescope you should recollimate it. That is not what I do. I leave my permanently stationed telescopes for many months on end without even checking them. In my defence, I find that it is extremely rare for me to have to make even the slightest of adjustments.

It all depends what materials the telescope is made from and how robust it is. For instance, a telescope home-made from wood is likely to need frequent adjustments owing to the unstable nature of the material.

Of course, it is important to check any portable equipment each time it is set up anew. So, if you are forced to use portable equipment then checking the collimation of your telescope becomes a necessarily frequent chore.

The following notes, concerned with the main types of telescope, may be of help. In all cases I must leave you to become familiar with the types of adjustments provided (push-pull screws, nuts and springs, nuts and lock nuts) and their locations on your own telescope.

Collimating a Newtonian reflector of focal ratio $f/6$ or larger

The first step is to make or buy a 'dummy eyepiece'. This is really no more than a plug which fits into the telescope drawtube and which has a small hole drilled exactly centre in the top of it. The best size for the hole is about 2 mm. This plug is inserted into the drawtube, replacing the eyepiece, and the function of the small hole is to ensure that your eye is steered onto the axis of the drawtube. When the collimation is successfully completed this axis will also coincide with the optical axis of the telescope.

You could make the 'dummy eyepiece' from an old 35 mm film canister but do make sure that the hole you drill in its base is exactly centred. Alternatively if you have an old high-power eyepiece that is no longer used you could remove its lenses and use that.

We will begin by assuming that the axis of the drawtube is exactly perpendicular to the side wall of the telescope tube. It certainly should be if the telescope has been commercially manufactured. How to check for this and correct any error is covered later.

Start by pointing the telescope at the daytime sky or at a light coloured wall, or an illuminated screen, wall, or curtained window if you are working at night. The secondary mirror mounting should have some provision for adjustments that will allow it to rotate and to move laterally up and down the axis of the telescope tube. Use these adjustments until you see, when looking through the 'dummy eyepiece', that the outer edge of the secondary mirror appears concentric with the inner edge of the bottom of the drawtube. The view you see should look rather like that illustrated in Figure 3.1(a).

If you rack out the drawtube this will make the secondary mirror appear to nearly fill your view of the bottom of the drawtube and so will help you to be more precise in your judgement.

Once you have successfully got the secondary mirror looking concentric with the drawtube you can make any fine adjustments to the tilt and the rotation of the secondary mirror cell until you see the reflection of the primary mirror appearing concentric within the inner edge of the secondary mirror (see Figure 3.1(b)). As before you can rack the drawtube in or out in order to get the reflection of the primary mirror nearly filling the secondary mirror. Even small inaccuracies will then easily show up.

All that is left is to adjust the tilt of the primary mirror cell until the reflection of the secondary mirror is precisely centred within it (as in Figure 3.1(c)). I always find the secondary support vanes of help here. Any slight error in the tilt of the primary mirror makes the reflections of the vanes very obviously unequal, as seen through the 'dummy eyepiece'.

You could visually check through everything again, refining your adjustments if necessary. Your telescope is now adequately collimated.

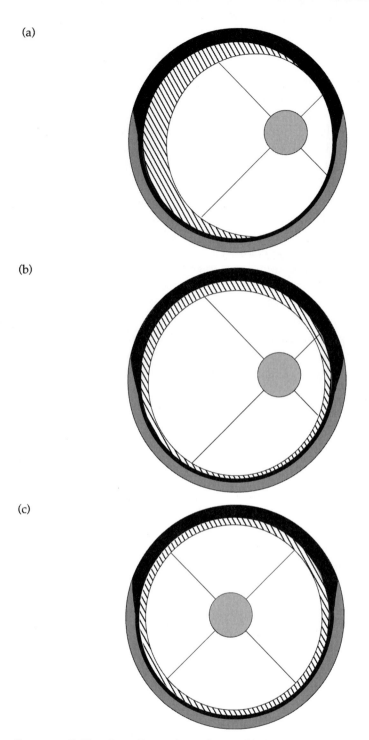

Figure 3.1 Collimating a Newtonian reflector: (a) the view through the 'dummy eyepiece' after adjusting the position of the secondary mirror to make its visible edge concentric with the edge of the drawtube; (b) the view after fine adjusting the tilt and rotation of the secondary mirror to make the reflection of the primary mirror concentric within it; (c) the view after adjusting the tilt of the primary mirror to make the reflection of the secondary mirror-mount concentric.

3.2 Collimation

Collimating a Newtonian reflector of focal ratio less than $f/6$

The lower the focal ratio of the primary mirror, the more critical is the collimation of the telescope. So, if we are to get the best results, we should use a refined technique to achieve the best possible collimation.

Added to the foregoing, there is another slight complication we encounter with low focal ratio Newtonian reflectors. We can collimate our telescope with the secondary mirror concentric, as before. However, the area of the field of view where the image is unvignetted (fully illuminated by all the rays arriving from the primary mirror) will be a little offset from the centre of the field of view.

To counter this, the secondary mirror ought to be offset a little away from the eyepiece and an equal distance towards the primary mirror. However, with focal ratios of more than $f/3.5$ for a 300 mm aperture, more than $f/4$ for a 400 mm aperture, and more than $f/5$ for a 500 mm aperture the adjustments will be less than 7 mm in each direction. The fall-off in image brightness due to the vignetting will be quite small, especially if the secondary mirror is large enough to illuminate fully a patch at least a couple of centimetres across (see Section 2.4). In that case, if you have any provision for making the offset adjustment (or if you are building your own telescope, or modifying it yourself to minimise outfield vignetting) I would say do not bother offsetting.

However, if you are using a commercial large reflector of low focal ratio, then the manufacturer may well have set the secondary mirror off-centre within the telescope tube, leaving you with no choice. You must ensure that the secondary mirror is also offset the correct distance down the axis of the telescope tube in order that the axial rays from the centre of the primary mirror are turned through exactly 90° and passed exactly along the axis of the drawtube.

Of course, you will realise that the rays can still be turned through 90° with the secondary offset away from the eyepiece but not equally offset down the telescope tube. This simply depends on having the secondary mirror set at exactly 45° to the optical axis of the primary mirror. *However, the normal collimation procedure will produce an error in that case: the mirrors will appear collimated to you but only when the secondary is actually tilted by more than 90°. The primary mirror will also be incorrectly tilted and the end result will be that the optical axis will pass at an angle through the drawtube.* Hence the need for the dual offset.

So, carefully measure the position of the secondary mirror to see if it is centred in the telescope tube, or if it is offset and by how much. If it is offset, then you must ensure that the mirror is offset by an equal distance down the telescope tube.

How do you do this? There are various ways. In my view the best method is by direct measurement. Measure the minor-axis diameter of the secondary mirror (so you can determine where its centre is), the diameter of the bottom end of the drawtube, and the distance this is from the sky end of the telescope tube (rack the focuser inwards so you can make the measurement). From this you can determine the distance of the exact centre of the axis of the drawtube from the top of the telescope tube. Putting a straight edge across the top of the telescope tube you can measure down to the top edge of the secondary mirror. When the mirror is correctly set at 45° its centre will be this distance plus the minor-axis radius from the top of the telescope tube.

If you want the secondary mirror centred with the axis of the drawtube simply make the distances from the top of the telescope tube to the centre of the secondary mirror, and to the centre of the drawtube, exactly the same. If you want, say, a 4 mm offset then move the secondary mirror a further 4 mm down towards the primary mirror. *Remember, the secondary's offset towards the primary mirror must be equal to the measured offset in the direction away from the eyepiece.*

Having sorted out the correct positioning of the secondary mirror we can now make the, hopefully very small, adjustments to bring the telescope into proper collimation. We have a choice. We can, in daylight, run through the same procedure as outlined in the last section and then take our telescope out under the stars and check for any remaining inaccuracy based on what we see through the eyepiece and make the very fine adjustments that may be needed to finish the job. The second option is to first run through the same procedure as before and then, still in daylight or using an illuminated screen or wall, apply another piece of kit to bring the collimation to a higher degree of accuracy.

The first option may sound OK but making mechanical adjustments to the telescope at night is tricky. Also you want to be observing, rather than fiddling with the telescope at this time (unless, of course, it is a portable telescope set up at the beginning of your observing session in which case you have no choice). In the next section I cover star-testing a Cassegrain telescope and refining the alignments this way. You can follow the same procedure for fine-tuning your Newtonian telescope.

The second option is usually much preferable but does require you to have placed a mark on the primary mirror (something you may feel nervous about doing – but you may be lucky and have a mirror the manufacturer has centre-marked for you). The following paragraphs describe the procedure.

Carefully measure the exact centre of your telescope's primary mirror and mark a small spot at this point. It has to be correctly placed to ±1 mm (albeit you are relying on the manufacturer getting the optical centre of the mirror at this same point). A spirit marker will suffice to make the mark. *Do please keep your fingers, palms, sleeves, and cuffs away from the mirror surface. As well as the obvious hazards of scratches and dust, the oils, moisture, and salt from your skin will badly affect the reflective coating.*

While you are taking on this invasive procedure you might as well check to see that the primary mirror is properly centred within the telescope tube to better than ±2 mm. If it is not and there are no provisions for adjustment then all you can do is to bear this fact in mind when deciding about any offsets for the secondary mirror. Improve the centring of the primary mirror if there is any way of doing it – perhaps there are provided radial adjusting screws or bolts. You

3.2 Collimation

might consider using shims but do not unduly squeeze the glass of the mirror or you will see the star images distorted as a result.

Next go through the collimation procedure as described in the last section. You can be fairly quick about it this time and no 'dummy eyepiece' will be needed for this rough set-up. Then insert a carefully made 'sighting tube'. This is a tube about 18 cm long which fits snugly into the drawtube. One end is closed with an accurately made 'dummy eyepiece'. The other end (this end is inserted into the drawtube) is open but is crossed with wires. The wires are mutually at right-angles and cross at the exact centre of the tube. 'Sighting tubes' can be purchased commercially. If you make your own, you must ensure that the sight hole and the crosswires are accurately positioned.

Using the 'sighting tube', make any fine adjustments to the secondary mirror's tilt and rotation so that the reflection of the spot on the primary mirror appears exactly under the intersection of the crosswires.

For the next adjustment you may need to use a lamp to throw some light into the telescope tube so the crosswires are illuminated enough to be seen via reflection in the telescope mirrors. Fine adjust the tilt of the primary mirror until the reflection of the crosswires you see exactly coincides with the ones at the end of the 'sighting tube' (and so the intersection of the reflected crosswires also coincides with the centre spot on the primary mirror). Finally check through the procedure again. If you were to star-test your telescope now any viewed errors remaining would likely be the fault of the optical manufacturer.

Collimating a Cassegrain reflector

As before, it is a good idea to check that both the primary and the secondary mirrors are properly centred within the telescope tube. It is true that both could be off-centre, even by different amounts, as long as both mirrors can be tilted enough to bring their optical axes into coincidence. However, in that case the focuser drawtube would also have to be titled by an amount and direction to suit. If the two mirrors are properly concentric to start with the drawtube can then be perpendicular to the base-plate and all will be well (assuming it is also mounted properly centred on the base-plate). Cassegrain telescopes are often provided with adjusters for this purpose. If so, do your best to make all the components concentric with the telescope tube to ±1 mm.

Next point the telescope at a light coloured wall and insert a 'sighting tube' into the drawtube. Looking through it adjust the tilt of the secondary mirror until the reflection of the primary mirror in it appears exactly centred. Then adjust the primary mirror until the reflection of the secondary mirror in it is also exactly centred.

If the secondary support vanes are four in number and are equally spaced (as will normally be the case), then you can use the crosswires on the sighting tube to achieve a little extra precision. Simply rotate the 'sighting tube' until the crosswires are brought into near coincidence with the secondary support vanes.

With careful adjustment you should be able to get the crosswires to exactly coincide with the reflection of the support vanes.

Now your telescope will be very close to its optimal collimation. Nonetheless, a Cassegrain reflector is rather more finicky than most other types of telescope and it is as well to be prepared for some *very* slight further adjustments when you actually try it out under a clear sky.

Here is how to go about the final refinements to the collimation. Plug in an eyepiece that will deliver a magnification of several hundred after pointing the telescope at a test star. The star ought to have as high an altitude as possible so that it is not too badly affected by atmospheric turbulence. If the telescope drive is rather erratic (as many are) you should choose the star Polaris for this test.

Once the test star is centred, defocus it slightly while watching for any asymmetry. If the expanding disk of light becomes oval and is less bright in one direction, the telescope is still very slightly misaligned. You should be aware that this test assumes the optics are of excellent quality. If the mirrors are even slightly astigmatic then the images will display a distortion which will be rather hard to tell from misalignment.

Now comes the really tricky bit – made easier if you have a willing assistant. With the telescope still pointed at the test star adjust the tilt of the secondary mirror until the star moves a little in the direction in which the out-of-focus image is at its faintest and most distended. This will be an adjustment so slight that tightening the one screw without slackening the other two may even suffice, depending on the robustness of the mounting of the secondary. Next adjust the tilt of the primary mirror by just enough to bring the star image back into the centre of the field of view. For this to work the telescope must not have been joggled out of its alignment with the test star while the adjustment to the secondary was made – an incredibly difficult thing to achieve in practice.

If all has gone well the slightly out-of-focus star disk will now appear rather more circular and evenly illuminated. Continue this tricky procedure until you achieve the most circular and evenly illuminated star disk you can manage. Check that it remains so for all positions of the focuser.

Collimating a refractor

If the manufacturer has not provided any adjustments for the squaring on of the object glass, then it is still worth checking the alignment but you will have to either put up with any misalignments or return the instrument to the manufacturer. If any adjustments are provided then the procedure is rather similar to that for the fine tuning of the Cassegrain reflector.

Select a good 'test star' as before. Centre it in the field of a high-power eyepiece. Slightly defocus it and watch for any expanding asymmetry in the expanding disk of light. Try to ignore the different colours you will see as the disk of light expands. This is quite normal behaviour for a refractor's object glass. All you are interested in detecting is any non-circularity and uneven distribution of light.

Alter the tilt of the object glass until the recentred star image looks as symmetrical as possible both a little inside and a little outside the best focus position.

Collimating Maksutov and Schmidt–Cassegrain telescopes

Fortunately the modern ones always come with instructions on how to collimate them. These instruments usually have only the secondary mirror (in the case of the Schmidt–Cassegrain) or the corrector plate with the secondary mirror (in the case of the Maksutov, where the secondary mirror is an aluminised spot in the middle of the corrector plate) adjustable for collimation, when any provision for adjustment is provided at all. My advice is always to follow the manufacturer's instructions. If you have acquired an instrument without any instructions then, after inserting a 'dummy eyepiece' and altering the tilt of the secondary mirror until the reflection of the primary appears concentric within it, fine-tune as I have previously described for the Cassegrain reflector.

Collimating a Schmidt–Newtonian telescope

Meade's line of Schmidt–Newtonian telescopes come with full instructions on monitoring and adjusting the collimation. In brief, the procedure is the same as for a Newtonian telescope except that the secondary mirror is not readily accessible to you for adjustment. The Meade instruments come complete with a factory-made collimation spot at the centre of the primary mirror. The only optical adjustments you should normally make are to adjust the tilt of the primary mirror. It is a good idea to star-test and, if necessary, fine adjust the collimation in the field.

'Squaring on' the eyepiece focusing mount

If the drawtube is not aligned to the optical axis of the telescope, the focal planes of the eyepiece and the telescope will be tilted with respect to each other. The action of focusing the telescope can then only achieve coincidence of these planes along one line. Star images will be in sharp focus anywhere along this line but will become increasingly out-of-focus with increasing distance from it. This is a very serious error to have in any instrument intended for use in astrovariable work.

With Cassegrain, Schmidt–Cassegrain, and Maksutov telescopes drawtube alignment is easily checked using a 'sighting tube'. Look through it and check that the crosswires appear centred against the secondary mirror in each case. For checking a refractor make a cap for the object glass with a small hole drilled exactly central. Does the intersection of the crosswires coincide with this hole?

Mechanically checking a Newtonian telescope's focuser is more difficult. It involves removing the secondary mirror and cell and marking a small spot on the far wall of the telescope tube *exactly* opposite the centre of the drawtube. The position of this spot can only be determined by very careful measurement. With the 'sighting tube' inserted, do the crosswires line up exactly with this spot?

Focuser–drawtube misalignment is perhaps best checked by star-testing. Firstly, centre the test star and focus it as carefully as possible. Try watching as you rack the focuser from fully in to fully out. As the disk of light changes size does it also appear to shift across the field of view? If so, try to determine (by waggling the end of the drawtube if necessary) whether this is merely 'slop' in the sliding fit or if there is a progressive lateral shift of the image which can only be due to misalignment.

I should say here that commercial Maksutov and Schmidt–Cassegrain telescopes achieve focusing by moving the primary mirror slightly along the optical axis. You will experience a cyclical to-and-from motion of the image as you change focus due to the inevitable slight mechanical imperfections of the mechanism. Also it isn't easy to do anything about any drawtube misalignment if you do find any.

Investigate further by moving the telescope in order to place the star at different positions in the field of view. Do you notice any change in the appearance of the star as it is moved across the field of view? Do bear in mind that it will be normal to see some image degradation radially away from the centre of the field of view. This will be especially evident close to the edge of the field of view. However, is this degradation symmetrical about the centre of the field of view?

Cassegrain telescopes (now rarely sold to amateurs but second-hand ones do come on the market) often come with provision for adjusting the focuser alignment. Most other telescopes do not, so you will either have to put up with the defect or resort to packing the mounting of the focuser with shims in order to achieve proper alignment.

Collimating a 'star diagonal'

At one time scorned by many, star diagonals have now become almost mandatory with today's manufactured refractors, Maksutovs and Schmidt–Cassegrain telescopes. The best examples of these units have some provision for adjusting the tilt of the mirror/prism.

The test for proper alignment is simple but should only be made after the telescope is properly collimated (including the alignment of the drawtube). Centre a star in a high-power eyepiece (better still if it has fitted crosswires) and carefully focus. Check again that it really is properly centred. Replace the eyepiece with the 'star diagonal' and put in the same eyepiece. Refocus. Does the star still appear centred? If not adjust the tilt of the mirror/prism until swapping between diagonal plus eyepiece and eyepiece alone produces no apparent image shift.

Other aids to collimation

You do not really need anything more than I have described in the foregoing notes. However, there are some devices to further help you in collimating your telescope that you can purchase if you really want to.

3.2 Collimation

For instance there is the *Cheshire Eyepiece* and the *Autocollimating Eyepiece*. Their principles of operation are too involved for me to describe here but you can rest assured that their manufacturers always provide full instructions with them. If you want to know more about them, there is a good article in the March 1988 issue of *Sky & Telescope* magazine.

The *Laser Collimator* is a very popular device these days. It is inserted into the drawtube and produces a thin beam of laser light which is passed through the optical system of the telescope. The idea is that the secondary mirror is adjusted until the spot of laser light falls exactly onto the centre of the primary mirror (on the centre spot if there is one). The primary mirror tilt is then adjusted until the laser light passes back into the device, where some of it is passed through a partially reflective mirror and appears on a 'target' which you view from the side. Supposedly this means that the laser beam has passed back exactly along its outgoing path and so ensures that the telescope is properly collimated.

Unfortunately there are hazards. One is the slight but very real physical hazard of catching a dose of laser light in your eye. The second is that a drawtube misalignment, or a secondary mirror misalignment, can produce a situation whereby you adjust until the spot of light hits the target but the beam has *not* actually passed back along its outward path. You will then have actually set your telescope with a definite misalignment between the axes of the secondary and primary mirrors! Collimate your telescope as closely as you can using the procedures I describe in the foregoing notes and limit your use of the laser collimator to the final fine-tuning.

A better approach to using a laser collimator with a Newtonian reflector is described by Nils Olof Carlin in the January 2003 issue of *Sky & Telescope*. The first stages are carried out as previously described. His innovation is for fine adjusting the tilt of the primary mirror. For this final stage he suggests inserting the laser into a Barlow lens and plugging this into the telescope drawtube. The bottom of the Barlow lens is closed with a disk with a small central hole in it (to allow the diverging laser light to emerge into the telescope). Instead of a central spot on the primary mirror there is a small central ring (a self-sticky 'reinforcement ring' used normally for sheets of paper to go into a ring-file might do very well).

When the telescope is correctly collimated a silhouette, in laser light, of this small ring appears concentric with the hole in the disk at the bottom of the Barlow lens (if it is too far up inside the focuser for you to see it use a small mirror hand-held inside the mouth of the telescope tube). The clever thing about Carlin's innovation is that the broadness of the cone of light makes the system insensitive to secondary mirror offset and drawtube misalignment, while still being very sensitive to primary mirror misalignment. Hence the situation of an inaccurately offset secondary mirror is not made far worse by introducing a tilt to the primary mirror in order to bring the laser spot to target as it would in the usual way of using the device.

3.3 Finding your chosen variable

In Chapter 1 I discussed the finder charts you might obtain from your astronomical society/association. You already have several hundred available for your use on the CD-ROM that accompanies this book. The target object is normally positioned at the centre of the chart and the magnitudes of several suitable comparison stars are marked.

If you are observing a star of naked-eye brightness, then of course you should be able to locate the star with the simplest of aids, maybe just using a simple chart or atlas such as *Norton's Star Atlas* (edited by Ian Ridpath, twentieth edition, Pi Press, 2003).

Fainter stars need a more sophisticated approach in order to acquire them in the field of view of your telescope or binoculars. If you have a telescope which is computer controlled, then simply keying in the known co-ordinates of the star ought to be sufficient to enable the telescope to point to within a few arcminutes of the chosen target. Using your widest-field eyepiece, this should be sufficient for you to recognise the pattern of stars on the finder chart (taking into account the size of the visible field).

At least that will be the case when the telescope is accurate enough in its 'GO-TO' operation. If it is not, then maybe you can improve the accuracy of its initial set-up. This might be done by the careful levelling of an altazimuth mount, or a more accurate polar alignment of an equatorial mount, etc. The same goes if your telescope is manually operated but has *accurate* setting circles (a rarity for modern commercial instruments) or has computer-aided manual pointing.

If you are using binoculars or a manually operated telescope without accurate setting circles, then you will have to resort to a technique known as *star-hopping*. For this you will need a good quality star atlas that shows stars at least as faint as the brightest two or three of those shown on the finder chart. In fact, the fainter the atlas goes the better. It should also have the largest possible scale.

Wil Tirion's *Sky Atlas 2000.0* (edited by Wil Tirion and Roger W. Sinnott, second edition, Cambridge: Cambridge University Press, 1999) is an excellent atlas for use with binoculars or a telescope with a very wide-field eyepiece. It shows stars down to eighth magnitude. A total of 43 000 stars are plotted on 26 charts, the scale being about $1°.2$ per centimetre. A $1°$ field of view should contain at least one atlas star. This will be of borderline usefulness for star-hopping if you are using a typical telescope but will be ample if you are using an instrument with a field of view of $2°.5$ or more.

For telescopic use you will be better off with another atlas by Wil Tirion (with Barry Rappaport and George Lovi): *Uranometria 2000.0* (edited by Wil Tirion, second edition, Willmann-Bell, 2001). Nearly a third of a million stars, down to tenth magnitude, are plotted on this two-volume atlas. The 500 charts show the sky at an average scale of $0°.6$ and there is a very convenient grid superposed.

3.3 Finding your chosen variable

Of course, there is now a plethora of atlas-generating computer software (such as *Starry Night Pro*) which you can use and so print-out a star map of your own. Some have the facility to set the faintest magnitude of the stars plotted and the print-outs can be whatever scale you choose. Consult magazine advertisements and reviews to see what is available and good value for money at the time you need to go shopping.

When star-hopping I find the quickest and easiest technique is first to inscribe on a clear overlay sheet (to avoid marking my expensive atlas) a circle which represents the size of the field of view of the eyepiece at the scale of the map/printout/atlas. Then I set the telescope to the nearest bright star to the location of the variable and begin by placing the acetate sheet so that the star lies at the centre of the circle. Next I move the overlay so that the circle moves in the general direction of the variable but, importantly, moves onto a recognisable pattern of stars. The move is less than one field diameter, so that the new field at least slightly overlaps the first. I then move the telescope until I recognise the new star pattern in the eyepiece. I continue the progress until the circle, and then the telescope's field of view, is centred on the variable star.

Of course, if the map is simply a computer print-out (or a photocopy from the atlas) then the pencilled circles can be compass-drawn straight onto it. In this case a series of overlapping circles leads to the variable star. Actually, this is a better way of doing it as the marked map can be kept for use on all future occasions.

Believe me, if you try to locate your chosen variable for the first time (using a manually slewed telescope) without star-hopping you will take a very long time to do it and you may even confidently alight on the wrong star! Please don't take short cuts. Be meticulous. After the first few times you will find that you can use this technique to set the telescope on the correct star very quickly. Without star-hopping locating on the variable will remain a lengthy, frustrating, and chancy procedure.

Of course, the finder chart will enable you to make the final identification of the variable star but do not be thrown if it proves to be unexpectedly bright or, more commonly, unexpectedly faint. If it is a variable of large amplitude you might even find that it is invisible if you have caught it at a time of deep minimum brightness.

Let me repeat, your first attempt to set the telescope on a given variable star may well take you a long time. Do not be put off. On the next occasion it will be a little easier. Soon setting on that particular star will become a part of your repertoire and finding it will take you no more than a minute.

Once you have the star located in the field of view, having used the full aperture of your telescope, you can select the magnification, and maybe the appropriate aperture stop, in order to bring the variable to an acceptable brightness (ideally two or three magnitudes brighter than the faintest star you could detect in the field of view) and with the correct size of the field of view in order to see the variable together with its comparison stars all occupying the central two-thirds to three-quarters of the field or less. See the previous chapter for discussions about eyepieces and the limiting magnitudes you can expect with given apertures. We are now, at long last, ready to make the actual brightness determinations.

3.4 Making the magnitude estimate

Now that we have our chosen astrovariable in view, we can undertake to determine its current brightness. There are a number of techniques we can use to do this. There are also some variations of these that some observers have 'customised' for themselves. With practice you should be able to get results consistent to yourself accurate to $\pm 0^m.1$, in all but the most difficult cases.

That is not quite the same thing as saying that the results are truly that accurate against those that would be obtained by accurate photometric techniques. You might consistently register the star as being a little brighter or a little dimmer than it really is. Even if not you will certainly record the brightness of the star with random errors amounting to plus or minus $0^m.1$ at the very least.

However, this is where the pooling of your results with those of others greatly enhances the value of your work. The mean of all the observers' values gives a truer measure of the actual magnitude of the variable star and the scatter in the combined results of the contributing observers gives a measure of the uncertainty in that value. The resulting light-curves generated from the pooled results will also allow a much more detailed and accurate examination of the variations of the star's brightness with time. This analysis will include testing for, and the determination of, any component periods of oscillation. Any one observer's results of a particular star will quite likely contain too many gaps, and have too great a scatter due to errors, for meaningful analysis.

I will outline the three main basic techniques for making variable star visual estimates in the following notes. Personally, I find that the second method normally suits me best but there is a certain degree of overlap between them. Try them all for yourself and give yourself time to decide which works best for you. Then practise that technique until you are proficient at it.

Comparison star sequence method

This is the method most observers use to estimate variable star brightnesses. After first looking at the variable star in order to gauge its brilliance, search the eyepiece field in order to find a star of identical brilliance. If successful the variable is, of course, assigned the same magnitude value as the comparison star.

Normally no such identical comparison star is found. In that case, search the field to find two stars of as near as possible the same brilliance as the variable. Ideally, one star should be a little brighter than the variable and one a little fainter. As an example, if the comparison stars have magnitudes $8^m.3$ and $8^m.7$ and the variable star is exactly half-way between the two then we can assign a magnitude value of $8^m.5$ to it.

3.4 Making the magnitude estimate

Of course, we would be lucky, indeed, to find such an easy case. You might find in practice that the star is closer in brightness to the $8^m.3$ star, so you would assign a value of $8^m.4$ to the variable. Even more likely is that the comparison stars do not bracket the variable so conveniently. There might be a slightly dimmer star present but the next brighter one might be very much brighter. In that case the best you can do is to estimate fractional differences of brightness and do your best to achieve an accuracy of $0^m.1$ in your determinations. This is where this method overlaps with the next technique.

Fractional method

As before we are comparing the brightness of the target variable with the brightnesses of comparison stars in the same field of view. We are looking for fractional differences in brightness and we record our observation in such a way that we can subsequently determine the magnitude value of our variable star after the observing session.

As an example, let us say that our variable's brightness lies between that of two comparison stars A and B. After careful scrutiny we decide that the variable's brightness is closer to that of star A, say two-fifths of the way from A to B. We should record the observation as: A 2 V 3 B.

After the observation we look up A's magnitude. Let us say that it is $9^m.1$. Let us suppose that B's magnitude is $9^m.6$. The difference between the two magnitude figures is $0^m.5$. Since we expressed the variable's brightness in terms of fifths of the way from the brightnesses of star A and star B we are interested in units of one fifth of $0^m.50$. Of course, this is $0^m.10$.

So, using star A:

$$\text{magnitude of variable} = 9^m.1 + (2 \times 0^m.10) = 9^m.3$$

As a check, we also use star B:

$$\text{magnitude of variable} = 9^m.6 - (3 \times 0^m.10) = 9^m.3$$

Do note that I have used two decimal places in the value for the magnitude difference, even though this is much smaller than the magnitude difference detectable by the human eye. The reason for this apparently spurious accuracy is to avoid any extra error being introduced caused by rounding the figures. For instance, suppose that another variable star has two convenient comparison stars of magnitudes $8^m.0$ and $8^m.7$ and we decide that the brightness of the variable is between that of the comparison stars and two-fifths of the way from that of the brighter comparison star. This time the two comparison stars are $0^m.7$ different in brightness and one-fifth of this magnitude is $0^m.14$. If we round this figure to $0^m.1$ before the calculation, the magnitude of the variable comes out as $8^m.2$. Leaving our rounding to the end of the calculation gives us the more accurate figure of $8^m.3$.

Pogson's step method

This time we compare the variable star directly to one comparison star, chosen to be as close as possible to the brightness of the variable. The observer simply estimates the number of 'steps' of brightness that separates the comparison star from the variable. The size of the steps you choose is your own responsibility but should be as close to $0^m.1$ as possible. This obviously means some prior practice in judging $0^m.1$ steps. You could get that experience by comparing non-variable stars before embarking on your variable star programme.

If the comparison star, let us call it A, is of magnitude $10^m.8$ and you judge the variable to be two steps brighter, then its magnitude must be $10^m.6$. You would initially record the observation as 'A + 2' and would look up A's magnitude after the observing session. When working out the magnitude of the variable, do remember that '+ 2 steps' is numerically equal to $-0^m.2$ (or adjusting this to your own chosen step size).

In essence, though not in practice, that is the end of the observation. Of course you should really check your observation by using at least one more comparison star. Let us call this one B. If you think the variable is three steps fainter than B, then you would record this as 'B −3'. If you find that the magnitude of B is $10^m.3$, then that puts the magnitude of the variable as $10^m.6$ in agreement with your first estimate. If your figures disagree by more than $0^m.1$ then discard the observation and go out again and repeat it. In any event you should really try to compare the variable with a third comparison star, if a suitable candidate exists.

3.5 Some difficulties and some remedies

Every variable star observer using visual means ought to strive to achieve an accuracy of $\pm 0^m.1$ in his/her measurements. However, even just discerning, let alone accurately estimating, brightness differences this small is no easy task.

Without looking up the magnitude values of the stars beforehand, take a look at the three belt stars of Orion. The westernmost star, Mintaka (δ Orionis) is very obviously fainter than the other two. Of the other two stars Alnitak (ζ Orionis), which is the easternmost star, and Alnilam (ε Orionis), one is $0^m.1$ brighter than the other. Can you tell which one is the brighter star? Decide for yourself before looking up the magnitudes in a catalogue.

Another difficulty arises when the variable star is a different colour to the comparison stars (and the comparison stars are themselves of differing colours). Many variable stars, particularly the type known as pulsating variables, are rather reddish in colour. Comparing the brightnesses of different coloured stars is not at all easy. At least this problem can be alleviated by having a good quality yellow filter screwed into the eyepiece when dealing with the most extreme cases.

Another pitfall associated with a star's colour is the *Purkinje effect*, which can badly distort your perception of the brightness of the reddest stars. There

3.5 Some difficulties and some remedies

are several aspects to the Purkinje effect. The worst of them is that the apparent brightness of a red star increases while you are looking at it. One way to mitigate this problem is to avoid taking long lingering looks at the star. You must, instead, decide on the magnitude of the star by taking short glances at it. As before, a yellow filter might help.

The brightness of the sky background can also affect the apparent brightness of a red star by a larger factor than for white or blue stars. Haze or moonlight tends to make a red star appear brighter than it really is when compared with others which are not so red. This is another manifestation of the Purkinje effect which can be alleviated by the use of a yellow filter.

I find that one way of increasing my sensitivity to brightness differences in stars is to defocus the stars slightly. I find that the small disks of light are much easier to compare than the sharply focused star images. As a bonus, this also virtually eliminates the Purkinje effect. The reason for this is that the Purkinje effect operates chiefly when a light source is concentrated into a small point.

Of course any defocusing will also cause the faintest stars that the telescope or binoculars can show to become invisible. In effect a defocused instrument will behave like one with reduced aperture, as far as establishing its limiting magnitude is concerned.

Sometimes a variable star will be too faint to observe with a pair of binoculars and yet it has no comparison stars of suitable brightness close enough to it in order for them all to be simultaneously visible in the smaller field of view of the telescope. In this case the best that can be done is to rapidly move the telescope between each comparison star and the variable, while trying to hold on to your impression of the brightnesses. Of course this is far from satisfactory and the error in the determined magnitude is likely to be several times larger than the desired for $\pm 0^m.1$. Thankfully, this situation does not arise too often.

Related to the last point, it might be that vignetting and/or off-axis aberrations in your telescope prevent brightness comparisons between stars variously placed in the field of view. If so, and you cannot change the instrument for another or make any modifications to cure this defect, then you will have to resort to sighting the variable and each comparison star one at a time at the centre of the field of view. Again, all you can do is try to remember the brightness of each. In that case you must make it clear on your submitted report that your results are less reliable because of this.

When making brightness estimates by naked eye you might find that suitable stars lie at inconveniently large distances away. At least you can quickly shift your gaze between the stars but you may well run into problems caused by variations of sky transparency and/or the level of background illumination. This is particularly the case for stars at significantly differing altitudes. The only remedy is to try to use comparison stars which are as close as possible to the variable star, even if they do differ in magnitude from it by more than you would like.

So, there is quite a lot to do in order to achieve accurate brightness determinations. However, after much practice you can (if you wish to) emulate observers like Gary Poyner and make many thousands of observations each year but let me repeat: *Just one observation carefully and accurately carried out is valuable. Fifty hasty and inaccurate observations are worse than useless as erroneous results can spoil any analysis, especially in the cases where there are few contributing observers.*

Chapter 4
Photometry

Most people who begin making visual brightness estimates of variable stars are soon hooked on it. In all probability you will be too. Also, you can be assured that for years to come the results of your current observational work will be useful to the astronomical community. At the time I write these words the majority of variable stars observers still do their work by making visual brightness estimates.

However, there will come a time, possibly not that many years away, when making further visual estimates of star brightnesses will no longer be considered to be scientifically worthwhile. The ubiquitous *charge-coupled device* (*CCD*) has already revolutionised astronomical imaging, first in the professional astronomical community and then in the amateur community.

In the meantime, the past and present visual estimates undertaken by you and your fellow amateurs will remain forever useful because they will likely be the *only* records of any particular star's brightness variations to date. When the day arrives that the frequent monitoring of the star is taken over by an accurate measuring device then, and only then, will it be no longer useful to make visual brightness estimates. That day has not yet arrived.

However, it is undeniable that even now the maximum value would be assigned to your observations if you could use an accurate and entirely objective method of determining stellar magnitudes. If you have, or can afford, a computer, a telescope with an accurate drive, and a CCD astrocamera, then you can undertake such precision *photometry* yourself.

4.1 Some basic principles of CCD astrocameras

A CCD consists of an array of light-collecting units, called *pixels*. Each pixel on the CCD has the same size as its neighbours. That size can range from about 7 μm to about 25 μm (7 micrometres to 25 micrometres) square. Professional astronomers mostly use large CCDs, typically having 2048 × 2048 or more pixels each around 25 μm square. These are very expensive. Currently amateurs use smaller versions. At the time I am writing these words a typical amateur's CCD might have an array of something like 500 × 750 pixels each perhaps 9 μm square. In that case the imaging area of the CCD would be 4.50 mm × 6.75 mm.

Whatever the size of CCD, the array of pixels is mounted on an 'integrated circuit' or 'silicon chip' type base which has about 20 individual electrical connections to its supporting electronics. The way it works is that photons of light falling on particular pixels liberate electrical charges within each of them. The more light (and so more photons) falling on a given pixel, the more electrical charge is created within it. If an image is focused on the picture receiving area of the CCD the pixels corresponding to the brightest parts of the image have the greatest amounts of charge liberated in them. The dimmest parts of the image generate the smallest amount of charges in the corresponding pixels.

Charges continue to build up all the while the light is falling, until each and every pixel is full, or *saturated*. Well before this stage is reached, the process, known as *integration*, has to be stopped. Ideally an *integration time* (equivalent to the photographic 'exposure length') is selected so that at the end of it the pixels associated with the dimmest parts of the image have only a small charge while those associated with the brightest parts of the image have lots of charge, though less than the amount necessary for saturation.

When the integration is completed the array of charges is sequentially read off the chip and sent as a representative data stream to a computer, or other electronics, to deal with in order to recreate the image on a monitor/TV screen, or to download it into a computer's memory, or onto a computer disk, etc.

A CCD is not equally sensitive to all wavelengths of light. The percentage of the number of photons of light falling on the CCD at any given wavelength that is detected by it is known as the *DQE* of the CCD at that wavelength. DQE stands for *detector quantum efficiency*. A DQE of 100 per cent is the best that one could possibly have; all the incoming photons then being detected.

Early CCDs tended to have their maximum sensitivity in the red, or even the near infrared, portion of the spectrum. A typical response might be a DQE of about 40–80 per cent in the 600–950 nm wavelength range falling away steeply at both shorter and longer wavelengths, to become zero at about 400 nm and again at about 1100 nm. This is very different to the spectral response of the eye, the maximum response of which occurs at a wavelength of about 550 nm in the yellow portion of the spectrum and falls to zero at about 380 nm (violet) and at about 700 nm (deep red).

Many of the more recent generations of CCDs have coatings which enhance their response to light at the blue end of the spectrum. As an example, the Philips FT12 has a response which is closer to that of the human eye. It has a peak sensitivity at 530 nm, falling to half that value at about 400 nm and 700 nm. However, it does so at the expense of some of its sensitivity, having a peak value of DQE of only 30 per cent.

4.1 Some basic principles of CCD astrocameras

So much for the basic principles. There are a number of variations in the design of modern CCD detectors. Look at the literature and you will come across the terms *full-frame*, *interline transfer*, and *frame transfer*. These refer to the way the CCD is structured and, consequently, how the image is read from the chip. You will also come across *back-illuminated* and *front-illuminated* CCDs. These terms refer to the mechanical structure of the CCD. Each type has its theoretical advantages and disadvantages (mainly in sensitivity, freedom from 'noise', resolution, and spectral response). You can do photometry with any of these types. In the real world, rather than in the theoretical realm, considerations of cost and pixel size and imaging area are of the greatest importance to you if you are to equip yourself on a limited budget.

One thing you should avoid in any CCD camera you buy for the purpose of doing photometry is what is called an *antiblooming drain*. As light falls on a given pixel the charges liberated in it will continue to accumulate. In a pixel with a built in antiblooming drain when the charges build up towards a certain level the excess charges are progressively bled off into the substrate of the CCD. The antiblooming drain is designed to avoid line overloads when photographing bright objects against dark backgrounds (such as street lamps in a night-time scene). Figure 4.1 shows the effect of overexposing a star using a CCD without an antiblooming drain.

You might think that an antiblooming drain would have nothing but advantages. In actual fact, the antiblooming drain exacts the twin penalty of making the CCD less sensitive and making its response non-linear. In other words, twice the intensity of light falling on the pixel does not produce twice the liberated charge in the pixel. This is a critical failing in a CCD to be used for photometry. This is important enough for me to emphasise the point: *Do not purchase a CCD with an antiblooming drain if you intend to do photometry with it.*

One problem afflicting CCDs of all types is something inaccurately called *dark current*. While an integration is underway thermally liberated charges build up in each of the pixels along with those liberated by the incident light. At room temperature these charges can build up to fully saturate each pixel in just a few seconds. Even before then the charges are reducing the total dynamic range (range of brightness levels) recordable. The effects are negligible for very short integration times, say a fraction of a second, but integrations longer than that demand the CCD be cooled. Practical CCD astrocameras have built-in thermo-electric coolers. Thermally generated charges work against the accuracy of your brightness measures and so any CCD used for photometry really ought to have cooling.

Limiting ourselves to the requirements of the photometrist, the basic characteristics we need to worry about in selecting a suitable CCD astrocamera are: cost, the size of the *imaging* area of the CCD (one-half of the area of the frame-transfer CCD is for imaging, the other acts as a storage area of the charges before they are read off), and resolution in the image (number of pixels height × number of pixels width in the image).

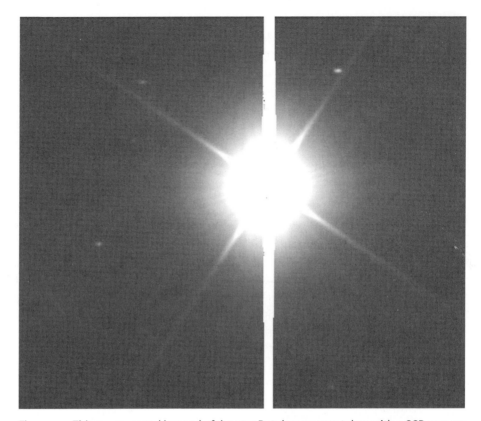

Figure 4.1 This overexposed imaged of the star Betelgeuse was taken with a CCD camera without an antiblooming drain. The charge from the star's image has overflowed to fill an entire column of pixels. The other spikes extending from the star are caused by diffraction effects from the telescope's secondary mirror support vanes. Image courtesy Nick James.

There is one other requirement. The whole system of camera plus computer (specifically its graphics card) must be capable of *16-bit* imaging. The term '16-bit' is computer jargon and relates to the number of separate brightness levels recordable. We need the 65 535 levels of brightness distinguishable by each pixel in 16-bit imaging if our photometry is going to have at least a potential accuracy of $\pm 0^m.01$. Systems having 12-bit (4096 brightness levels spanning black to white) or 8-bit (256 brightness levels) imaging fall well short of our photometric requirements, even if they do produce acceptable pictures.

At the time I am writing these words, most digital cameras and webcams do not have 16-bit greyscale imaging. The '24-bit' imaging you will often see advertised for them is actually three lots of 8-bit imaging; 256 brightness levels in each of red, green, and blue. Perhaps by the time you are reading this digital cameras, and even webcams, will be available with 16-bit greyscale imaging. If so, these devices can be used for photometry (albeit with the need for stacking lots of short exposures to synthesise a long exposure since they are not equipped with cooling). Most digital cameras also have antiblooming drains, which is a problem for our purposes. So, at the time I am writing these words we need a 'proper' CCD astrocamera to undertake our photometry.

4.2 The imaging area and resolution of a CCD camera when used on your telescope

As stated in Chapter 2, the image scale (in arcseconds per millimetre) at the focal plane of your telescope is given by:

$$\text{image scale} = \frac{206265}{f}$$

where f is the *effective focal length* of the telescope, taking into account the effect of the amplifying secondary mirror of a compound telescope and any other additional optics (Barlow lenses, etc.) in the light path before the focal plane.

When the CCD is placed at this focal plane each pixel will cover a number of arcseconds of the image. Knowing this figure and the size of the CCD's imaging area you can also calculate the actual size of the patch of sky the CCD is going to record in one exposure.

Let us take an example. You have a standard commercial 200mm $f/10$ Schmidt–Cassegrain reflector. Its focal length is 2000 mm. Using the foregoing equation you can calculate that the image scale at the focal plane is 103 arcseconds per millimetre. The camera you have is the SBIG model ST-5C. This camera has an array of 240 × 320 pixels, each 10 μm square. Hence 100 of the camera's pixels span 1 mm. Remember, this is also 103 arcseconds of image. Hence each pixel spans 1.03 arcseconds and the total imaging area of the CCD when used with that telescope (and no additional optics) is 247 arcseconds × 330 arcseconds.

The foregoing step-by-step example calculation can be summarised by the following general equations which you might find useful to have to hand:

$$\text{number of arcseconds of image per pixel on CCD} = \frac{206265 d}{1000 f}$$

where d = size of one pixel measured in microns (micrometres, μm) and f is the effective focal length of the telescope measured in millimetres.

$$\text{width of the image (in arcseconds) on the CCD} = \frac{206265 X \cdot d}{1000 f}$$

$$\text{height of the image (in arcseconds) on the CCD} = \frac{206265 Y \cdot d}{1000 f}$$

where X and Y are the numbers of pixels that make up the width and height of the CCD, respectively. If the CCD pixels are not square (they usually are in modern astrocameras) then the appropriate values of d have to be used to suit, of course.

When selecting a CCD astrocamera to be used for photometry with your telescope you should aim to achieve about 1–1.5 arcseconds of image per pixel. This means that an ideal camera to be used with a telescope of effective focal length 2000 m would have pixels sized about 10–15 μm square. This is not hard-and-fast but the best results would come with something close to this value.

I have already mentioned SBIG's ST-5C camera but the ST-7E and ST-8E cameras would also be reasonable choices. They have 510 × 765 and 1020 × 1530 arrays of 9 μm pixels. In these two cases each pixel would span 0.928 arcseconds and the CCD would cover 473 × 710 arcseconds of sky (for the ST-7E) and 947 × 1420 arcseconds of sky (for the ST-8E) when used with the same telescope.

Even 1420 arcseconds is only 23.7 arcminutes, or about $0°.4$ of sky. It is desirable to have enough sky area imaged so that comparison stars can be imaged at the same time as the astrovariable. Otherwise you will have to take separate exposures, one aligned on the astrovariable and one on each of the comparison stars. Unfortunately, the penalty is that larger chips cost very much more money. At the time I am writing these words you can purchase the ST-5E for just under $900 but the ST-7E will cost you almost $2700. The ST-8E is priced just a shade under $6000, more than double the cost of the 200 mm Schmidt–Cassegrain telescope you are likely to be using it with. Now you can see why most amateurs still estimate brightnesses by eye!

Why do I recommend imaging at a scale of about 1 arcsecond per pixel? The reason is that the front-surface CCD you are likely to be using has a mechanical layout that includes a series of strips laid between the rows of pixels. These strips incorporate the electrical gates which are a necessary part of the storage and readout mechanism of the CCD. These do not take part in recording light. If the star images are small, then a given star might be partly hidden by the gate structure. Only some of its light would be recorded by the CCD and subsequent analysis would show the star as being much fainter than it really is. However, if the image fell on the same CCD slightly differently then all the light from that star might be recorded.

The average diameter of the images of stars your telescope will deliver in normal seeing conditions is between two and four arcseconds when using an integration time longer than a few seconds (when atmospheric turbulence would have moved the star image around enough to produce a *seeing disk* a few arcseconds across). Thus most of the light from a star will be spread over a patch covering between about four and sixteen pixels. In that case all the star images will be similarly affected by masking due to the CCD's gate structure and so the desired comparisons of brightness can be safely made. For a similar reason, the more complicated gate structure and mask and microlens arrangements of colour CCD cameras (this is another problem arising with webcams and digital cameras) are even less suitable for photometry and should be avoided for all but their intended purpose of 'one-shot' colour imaging.

Incidentally, back-illuminated CCDs do not suffer from this problem as the light does not have to pass through the gate structure. In that case an image scale of two arcseconds per pixel is probably best in average seeing conditions. This creates a larger total field of view covered by the CCD. The rub is that back-illuminated CCDs are very much more expensive and need more careful handling in the field. For instance, they must be allowed to slowly warm up before deactivating – adding twenty minutes or so to the decommissioning time at the end of an observing run.

In more turbulent seeing conditions, but more particularly if you deliberately *slightly* defocus the image (so that star images are spread out into larger disks), you can obtain viable measures using a front-illuminated CCD at about 2 arcseconds per pixel, and so gain some extra field of view that way.

Accessories for Schmidt–Cassegrain telescopes (sometimes available for other telescope types, depending on the manufacturer) include *telecompressors*. These reduce the effective focal length of the telescope (and so they also reduce its effective focal ratio) allowing you to image more arcseconds per pixel on the CCD (if you need to – for our purposes they only become useful if the effective focal length of the telescope is greater than about 2 metres). They also increase the field of view up to a certain point. That point is less than you might imagine because of the effects of vignetting and optical aberrations. A common telecompressor offering $f/3.3$ when used on a typical Schmidt–Cassegrain telescope will illuminate a CCD of up to about a centimetre square with a good-quality image. The image would be too poor to utilise in the outer parts of a larger CCD. A telecompressor providing $f/6.3$ will do much better and might be a good compromise, especially if you are fortunate enough to own an astrocamera with a larger CCD.

Another way to effectively get more arcseconds per pixel is to use the camera in *binned* mode. For instance, a camera with a 500×750 array of 9 μm square pixels will behave like one with a 250×375 array of 18 μm square pixels when used in 2×2 binned mode. The field of view is the same, of course, but there are less resolution points in the image. One benefit is that the sensitivity of the system to star-like points is increased four-fold, provided the '18 μm pixels' still span less than about 2 arcseconds of sky. The sensitivity gains of telecompressing, when imaging star-like points, diminish if the image is undersampled (too many arcseconds per pixel). In any case, we really prefer sampling at the rate of not much more than 1 arcsecond per pixel (or effective pixel, if binned) for our photometric work.

4.3 CCD astrocameras in practice

Figure 4.2(a) shows the camera head of the Starlight Xpress SXL8 unit. Actually, I should say that this is one of the company's older cameras. However, many other companies' cameras look rather like this one which is why I have used this photo (Starlight Xpress Ltd have a website at http://www.starlight-xpress.co.uk on which you can view their latest products). Notice the cooling fins projecting from the back of the camera head. The major part of the mass of the camera head

Photometry

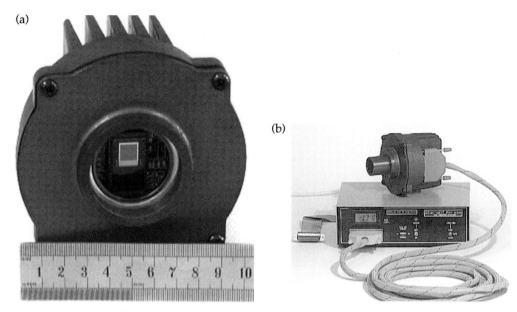

Figure 4.2 (a) The CCD can be seen within the camera head of this Starlight Xpress SXL8 unit. (b) The Starlight Xpress SX system, typical of commercial units. Courtesy Terry Platt and Starlight Xpress.

(about 1 kilogram) is associated with the cooling unit. The small grey square that lies within the head is the actual CCD. It is the Philips FT12, referred to earlier.

Figure 4.2(b) shows the rest of the Starlight Xpress SX system. Its mechanical construction is fairly typical of CCD astrocameras for the amateur market. The plug at the end of the ribbon cable is for attaching to the user port of the computer. Notice that the camera head has a short barrel fitted into the front of it. This is so the camera head can be plugged into the telescope drawtube (or adapter tube if one is using a Barlow lens, a relay lens, or an eyepiece to enlarge the primary image) in the same manner as one would plug in an eyepiece. In practice, the weight of the camera head necessitates rebalancing the telescope tube, perhaps using additional counterweights.

As well as the Starlight Xpress cameras, there are many others made by other companies. I have already mentioned some of the many units made by SBIG (their web-address is http://www.sbig.com). I recommend searching out advertisements in astronomy magazines current at the time you decide to purchase the camera. Get further information directly from the manufacturers and take the time to make your choice carefully.

When you have made your purchase read the manufacturer's instructions very carefully. The time and effort spent will be more than repaid by how quickly you will be able to achieve acceptable results. There will be variations in operation between one system and the next and so I will confine myself to offering general comments on matters relating to operating the camera with your telescope in order to perform photometry.

In use, you can expect the temperature of the CCD to stabilise in about 15 minutes after switching on the cooling unit. You must wait for this before attempting any photometric work. The temperature will be monitored and displayed by the supporting electronics unit as shown in Figure 4.2(b). Keep an eye on this reading during your observing run to ensure it really does stay constant – and discard your results if the temperature varies by more than a degree.

4.4 Getting the focused image onto the CCD and keeping it there

The goal is to have the telescope pointing at the target starfield and have a focused image, centred on the astrovariable, projected onto the CCD. Unfortunately, the small area of the CCD can make this difficult to achieve. If your telescope has a GOTO function then that is a great help provided it works accurately. A GOTO that places an object within an arcminute or two of the centre of the field of view is probably good enough if the CCD imaging area is at least 2 or 3 arcminutes wide and tall. Then you can use your finder chart to identify the object on your computer monitor and fine adjust to centre on it.

There is an accessory which can make acquiring the target very much easier: a *flip-mirror system*. Various companies such as Meade offer them. Their cost ranges from about $150 to $300. This small unit is inserted into the telescope drawtube just before the CCD camera. A small lever actuates a mirror. With the mirror 'up' it directs the light at right-angles into a viewing eyepiece. The telescope can be moved to place the subject at the centre of the wide field which is visible. Then the mirror is flipped 'down'. The light from the telescope is then free to pass into the CCD camera. The selected target will then appear on the computer monitor ready for fine centring.

If you are using a manually operated telescope, then I should say that a flip-mirror system is highly desirable if you are to avoid the frustration of spending a great deal of time trying to locate your target starfield. Another essential for a manually operated telescope is a good finderscope. It should be fitted with illuminated crosshairs.

One pitfall with the flip-mirror system is that it takes up several centimetres of focus. Will your telescope drawtube rack inwards far enough with the unit plugged into it to allow a focused image to form on the CCD? If not, you might have to be prepared to make alterations to the optical assembly (or maybe just install a new, lower-profile, focuser) to suit.

With a manually operated telescope, probably the best procedure is to plug in a wide-field eyepiece and find the required field using star-hopping, as I describe in Section 3.3. Centre on the target as carefully as you can. Next put in the flip-mirror system with the already attached CCD camera, trying to disturb the telescope as little as possible. Focus the image as seen in the eyepiece. With the flip-mirror 'up' centre on the target and then gently move the lever to lower the mirror and view the results on the computer monitor.

With the camera plugged into the telescope and the telescope trained on the target the next task is to focus the image. This is achieved by tweaking the focuser between exposures and monitoring the results on the computer. After doing this for the first time a mark made using a felt pen, etc. could be put on the drawtube, or focuser wheel etc., to speed things up in future. A motorised focuser would make life very much easier, allowing all the adjustments to be done by remote control while you are seated in front of the monitor. If you cannot afford one, perhaps you can make your own, or even motorise the existing one.

Of course, tweaking the focus, then performing an integration, waiting for the image to appear, then tweaking the focus again, etc. could certainly be time consuming. However, most astrocamera systems have a special 'focusing mode' whereby only a small area near the middle of the frame is imaged to speed up downloading the images. It produces a rapid sequence of images, allowing one quickly to achieve the sharpest possible focus.

Accurately training the telescope onto the target and getting to the point that the image is centred and focused on the CCD is not the end of the matter. We require that the image stays centred while we prepare everything for the exposure (you will have already noted I use the terms 'integration' and 'exposure' interchangeably) and we especially need the image to remain fixed in position on the CCD while the integration is taking place. If we are giving a series of exposures (as I recommend should be your normal procedure for each run on an astrovariable) we need the image to remain fixed during each of them and to remain sensibly so from one exposure to the next.

The amateur telescopes of yesteryear tended not to have very good tracking. With provision for slight adjustments in right ascension and declination and the use of a good guiding telescope, the best examples could keep the telescope on target well enough for recording long exposure images onto photographic film at the principal focus. Generally the tracking requirements are more exacting for imaging onto a small CCD, as you will be viewing the image at a greater enlargement factor on the computer screen or hard-copy print-out. However, it is also true that for photometry a few arcseconds drift is of no great disadvantage, even if the images do look cosmetically poor. Also of help is the fact that the maximum exposure we will need for our routine work is probably no more than a minute. Still, if your telescope refuses to track on its target by better than a few arcseconds during an integration you will have to adopt a strategy to overcome this defect.

Firstly, is there anything you can do to improve matters mechanical? If the telescope drive suffers from slop or backlash you might try adjusting the counterweighting, or adding new weights, in order to take up the slack in the system. An east–west jittery motion in the image, or even an erratic drift back then lurch forward effect, may well be cured this way. Many authorities recommend deliberately unbalancing the telescope in right ascension in order that the drive motor has to work harder to push the telescope round. I would add a note of

4.4 Getting the focused image onto the CCD

caution, though. Most amateur telescope drive motors are rather feeble. An overloaded synchronous motor might run half speed. It might even be permanently damaged.

Personally I think that unless the drive motor is unusually powerful it is better to unbalance the telescope so that it would slew to the west on its own if the drive gear was disengaged. In that case the drive would be acting as an escapement more than as a source of motive power. Even so, you could still overload the motor if you overdo the unbalancing to the point that the gear-train starts to bind up.

If you cannot improve your existing telescope mounting, one solution is to remount your telescope on a new one. Consult the current manufacturer and supplier advertisements in magazines, brochures, and websites and take time in making your choice. Be prepared, though, to spend a lot of money if you want a truly excellent mount such as those made by Losmandy, Software Bisque, or Takahashi.

Another alternative is to install an *adaptive optics* unit, such as that made by SBIG for use with several of their cameras. The current model is the AO-7 and it costs about $1300. The SBIG cameras have a second small CCD set next to the main one. The small CCD is for monitoring the tracking and results in signals sent to the adaptive optics unit. These cause tiny changes in the tilt of a mirror which results in the image remaining stationary on the main CCD despite any small tracking errors of the telescope. The system can correct up to about 2 arcminutes of tracking error when used with an effective focal length of 2 metres.

As with the flip-mirror system, the disadvantage is that several centimetres of focus are used up. In fact, with a flip-mirror system, an adaptive optics unit, and finally the CCD stacked and plugged into the telescope, you can be fairly sure that you will have to make a major adjustment to the positions of the telescope optics. An exception to this is the range of commercial Schmidt–Cassegrain telescopes: in these the primary mirror moves to effect focusing and there will be enough focusing latitude available.

Along with any changes you make, you will undoubtedly also have to take steps to reduce vignetting. For instance, you might have to change the existing secondary mirror of your Newtonian telescope for a larger one to cope with the wider cone of light it now has to intercept. Please refer back to Chapter 2 for more about vignetting and some other practical matters concerning telescopes.

The tracking headache is very much eased if you are lucky enough to own a modern telescope which is designed to be used with CCD cameras. Even so, you will still have to take note of the manufacturer's instructions to cope with factors such as periodic error in the drive gears.

One final possibility one can consider for correcting the effects of poor tracking is to *track-and-accumulate* exposures. This involves taking not just one exposure of the desired length but lots of short exposures and subsequently stacking the results via the image processing software. Each image is shifted enough that when they are stacked all the star images coincide. In that way a long exposure

Photometry

is synthesised from lots of short exposures. Twelve 5 second exposures might be selected as being well enough tracked for subsequent stacking from a run of twenty or more. When stacked they would synthesis a 1 minute exposure, though with a signal-to-noise ratio inferior to that of a genuine 1 minute exposure. Still, you would at least have a usable result with which to do photometry.

4.5 Taking the picture

Actually, our evening's observing run does not begin with taking the first images of our astrovariable. Our first task is to produce some *calibration frames* to use in the analysis after the observing run. How this is done is covered in the next section. For now I will continue with the train of operations we have been considering so far.

So, we will assume that at this point we have taken our calibration frames and we now have the focused image of the starfield, with the astrovariable centred, on the CCD. We switch from focus mode to normal imaging mode and so we next have to decide on the exposure.

The ideal result would be to have the pixels illuminated by the brightest part of the astrovariable, or the brightest comparison star, whichever is the brighter, about one half full of liberated charges (in technical jargon, these pixels are full to half of their *full-well capacity*).

The higher the 'count', the less it is afflicted by random 'noise'. Statistically, the uncertainty in the reading is inversely proportional to the square root of the count. So a count of 100 units of brightness has a 10 per cent uncertainty, while one of 10 000 units has an uncertainty of 1 per cent. This corresponds to about $\pm 0^m.01$ for that one brightness reading. For a count of 32 000 units statistics predicts an uncertainty of 0.56 per cent. It is desirable for each of the measures to have an uncertainty of *less* than 1 per cent if the final calculated magnitude is going to be accurate to $\pm 0^m.01$. Having the fullest pixels not much more than half full (say 35 000 units out of a possible 65 535 units) is desirable to avoid any inherent non-linearity in the CCD – and the fullest pixels most definitely should be rather less than half full if you have mistakenly purchased a CCD with an antiblooming drain.

When in doubt take a series of images, bracketing your best guess at the correct exposure. In fact, I would recommend taking multiple images every time. The longest integration times you will need probably will be no more than a minute. As a very rough idea, a 1 minute exposure will quite likely correctly expose a ninth magnitude star for photometry using a typical 200 mm aperture telescope and CCD camera. However, a great deal depends on the precise details of the equipment, the seeing conditions, and the sampling (how many arcseconds per pixel) on the CCD. There is also the colour of the astrovariable to be taken into account. In connection with that I should mention that you will need one or more filters in the optical train – but that is an involved subject which I will touch on later in this chapter.

The best advice I can give about determining the correct exposures is to get some practice with your system. Learn what exposures you require to produce images suitable for photometry with different integration times. You can predict that for every magnitude dimmer the star or astrovariable image is, you will have to increase the exposure by 2.5 times. Conversely, reduce the integration time to 0.4 (1/2.5) of the previous exposure for every one magnitude increase in star/astrovariable brightness.

4.6 Calibration frames

It would be great if one could quickly set up the CCD camera with the telescope, aim, focus, and take the exposure; then move a cursor about the image on the monitor to read off star brightnesses. Unfortunately things are nowhere near that simple. Instead we have to undertake an elaborate procedure in order to arrive at the data we need to calculate magnitudes.

The task begins with the making of calibration frames. How to make the required *flat field*, *bias frame*, and *dark frame* exposures will be fully described in the CCD manufacturer's instructions. These are all made with the camera plugged into the telescope in the same manner as for taking the actual images used for photometry. These will improve the cosmetic appearance of the recorded images. However, it is not for mere cosmetic reasons that we require these three operations. They are essential if we are to obtain meaningful and accurate brightness measurements from the image.

Flat field

A flat field is a short exposure of an evenly illuminated field of view. Its purpose is to eliminate the variations of image brightness that occur due to imperfections in the optical system. Specks of dust on the window will cast shadows on the CCD, for instance. Also vignetting will progressively darken the field of view towards the corners of the image. These imperfections and more can be accounted for in the processing that is done after a flat field is created.

Ideally a flat field is created by making an exposure of the twilight sky at the beginning of the evening's work. Do make sure that the camera cooling has been on long enough for the temperature to have stabilised before you begin. The sky ought to be sufficiently bright at the time that the correct exposure should be less than a second while most of the pixels are approaching half-full capacity (a brightness count of about 32 000, for instance). Do not exceed that 'half-full' level, though, because a few of the pixels will then be filled to saturation. The number of saturated pixels will increase with further exposure. Having a significant number of pixels saturated is undesirable in a flat field.

In the real world, most amateur astronomers have to contend with the vagaries of the weather and the myriad of other things in everyday life which mean that a 'twilight shoot' before the main session is seldom achievable.

An alternative is to set up a flat white screen in front of your telescope and illuminate it as evenly as possible. Some people lucky enough to have their telescope housed in a dome have used the inside of it as such a screen. Do this just before you embark on the evening's programme. You could (and probably should) add a further screen of white perspex or some other suitable material to the end of your telescope to diffuse the light further. Clearly you will need to be prepared to do some experimenting before you can settle on your preferred method for creating a flat field.

I mentioned that the exposure (integration time) for creating a flat field ought to be less than a second. This is to reduce the effects of the sources of 'noise' (particularly thermal noise) that would afflict a longer exposure. Noise shows up as graininess in an image and it represents an uncontrolled pixel by pixel variation that is highly detrimental to photometry. This is also the reason for having the light source bright enough such that most of the pixels are filled to the better part of half capacity (so that the ratio of signal to noise is favourable).

Even so, it is a very good idea actually to take a run of at least ten (twenty would be ideal) flat field exposures and later average the result of them (again, this is something that the supplied software will allow you to do). The final result will be a flat field with very little 'noise'. In other words the final flat field will be smooth (as opposed to grainy) and accurately mirror the variations in the optical illumination of the CCD by the telescope, just as we intend it should.

Bias frame

Many measuring instruments have a *non-zero offset*. In other words, when the quantity being measured is actually zero the measuring device does not show zero but instead indicates a small positive or negative reading. To get accurate measures one must always add or subtract this quantity, the *zero offset*, as appropriate from all the readings. CCD astrocameras also have a small non-zero offset inherent in the readout electronics. This is taken care of in the subsequent processing by utilising a bias frame.

To take a bias frame the telescope is capped (and/or the camera shutter is closed if there is one fitted) and the shortest possible exposure is given (this is probably one hundredth of a second, though a few camera operating systems do have provision for a 'zero' exposure length). As before take a series of at least ten bias frames to get an average. Do this just after you have completed the flat field run.

Dark frame

Dark frames are, like bias frames, taken with the camera or telescope shuttered against any light entering. This time, though, the exposures given are the same duration as those for taking the images. Hence raw dark frames are chiefly afflicted by thermal noise. When viewed such a frame gives a slight snowstorm appearance of 'hot' pixels against a grey background. As before, it is a good idea to take at least ten dark frames and use the average of them in the final processing. Mind you, this is time consuming if you need to image the chosen starfield with an integration time of more than a couple of minutes. Fortunately, integrations lasting no longer than a minute will prove to be sufficient in most of the cases we will encounter.

With our calibration exposures made we can then go on to take the image on which we will perform our photometry of the astrovariable and its comparison stars (as I described in the last section). Next I will describe part of the processing that will need to be done after the observing session is over.

Applying the calibration frames to the starfield image

The raw star image is afflicted with all the blemishes and faults that we have isolated in the calibration frames. The image processing software that will come with your camera (or which you can obtain separately if you do not wish to use the camera manufacturer's software) will have a number of functions, or operations, it can perform on the images you obtain with the camera. You will be able to subtract one frame from another, or add one frame to another, or you will be able to average frames, and even divide one frame by another. There are many other functions possible and commonly included in imaging software. I can only provide general notes here. It is down to you to study the literature supplied with the camera/software package and learn how to carry out these operations. In most cases this will involve nothing more frightening than clicking on options displayed in menus. You must spend the time and effort you will need to get to know the software package you are using.

For photometric purposes we will only need to average frames, subtract frames, and divide frames. Indeed, all the wonderful functions that can be carried out for image processing, such as contrast stretching, applying processing 'filters', image sharpening, etc. are most definitely to be ignored as using them would render our images useless for photometry.

The procedure is to bring up the raw starfield image and then subtract the averaged bias frame and then the averaged dark frame. The result of this is then divided by the averaged flat field. The resulting image is the one you should use for your photometry. For this to work well does require the temperature of the CCD be kept constant during the observing run, as I stated earlier.

Let me here restate a very important point: do not be tempted to improve the cosmetic appearance of the image beyond the application of the calibration frames. *Do anything further to the image and you will render it useless for photometry.*

It is difficult for me to show you an illustration of the calibration process necessary to render a raw image suitable for photometry because the on-screen visual effects are rather subtle. However, I can show you pretty much the same process in operation on an image to prepare it for *visual display*. See Figure 4.3. Here the contrast levels are much exaggerated and so these do show the effect of the dark frame and flat field well. Please, though, bear in mind that contrast-stretching is most definitely *not* to be used when preparing the image for photometry!

Photometry

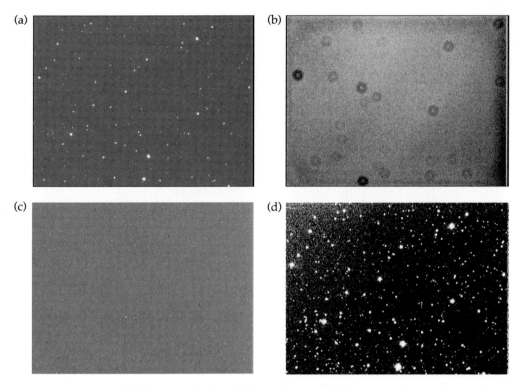

Figure 4.3 (a) Nick James obtained this raw image of HT Cas on 1995 December 9, using a Starlight Xpress CCD camera for a 3 minute exposure through his 300 mm $f/5$ Newtonian reflector. On this unprocessed image the sky background count is around 20 000 units on a scale of 0–65 353. It can be seen that the illumination of the field is non-uniform and some dust shadows are evident. (b) This frame is the flat field. It is an average of a number of short exposures of the twilight sky. Dust shadows are evident, as is a general variation of intensity across the CCD. (c) Even though the CCD is cooled it still suffers from thermal noise. This is the dark frame associated with the raw image. It is also a 3 minute exposure but was made with the cap over the end of the telescope. (d) The image having been processed for visual display (the dark frame subtracted from the raw image and the result divided by the flat field). In this case contrast-stretching has been used at various stages during the processing and the background level set at zero for the final image. These latter operations are NOT to be undertaken when preparing an image for photometry. Courtesy Nick James.

4.7 Obtaining magnitude measures from a CCD image

Thus far we have completed all of the really difficult tasks. We have our properly focused, properly exposed, and calibrated image on the computer screen. We save it in whatever format is compatible with the software package(s) we are using (commonly TIFF or FITS).

To finish the job (in all likelihood a day or so after our observing run) we have to bring up the image and apply the photometric functions contained in the software package we are using (probably the one supplied with the astrocamera).

4.7 Obtaining magnitude measures from a CCD image

The details of operation vary greatly from software package to software package. Most astrocameras come with software that includes the provisions for photometry. Other stand-alone photometric software packages also exist. At the time I write these words, the Windows-based AIP4WIN is very popular. It comes on a CD-ROM with Richard Berry's *Handbook of Image Processing* (published in 2000 by Willmann-Bell). Another stand-alone package is IRIS by Christian Buil. This one is currently free for download from his website at:

http://www.astrosurf.com/buil

Both of these programs are very powerful and can even automate most or all of the procedure.

Whatever the software we are using, we will be carrying out a procedure that is classified as both *aperture photometry* and *comparative photometry*. It is 'aperture photometry' because we use the software to define a small circle or square, the so-called 'aperture', around the astrovariable and each comparison star we are to measure. The software then reads the total amount of light recorded by the CCD within this small aperture. It is also 'comparative photometry' because the final determination of the brightness of the astrovariable is made by direct comparison with the brightnesses of the nearby stars of already known brightness.

As one example let me show you a screenshot of Christian Buil's IRIS program in action (see Figure 4.4). The inset shows a close-up of the star whose brightness is being measured. The inner circle contains the star image. It also includes a portion of the sky background. The outermost ring samples the sky background only. As you can see from the displayed window, the program makes all the calculations for you.

At the other end of the software market, so to speak, other programs generate just one aperture and leave you to select the size of this aperture manually, position it over the astrovariable and take a brightness reading, then position the aperture over each comparison star and record those brightness readings. Next you position the same aperture over one or more parts of the image which are obviously free of stars and again record the brightness levels. You then have to use a set of equations and your pocket calculator in order to get the final answer you want: the magnitude of the astrovariable.

As ever, it is for you to read and follow the instructions that go with the software package you are using. Here I can only highlight a few general matters that are important if your photometry is to be accurate.

To start with, this is the master equation you will need if you are using a manual photometry program:

$$\Delta m = -2.5 \operatorname{Log} \left[\frac{(C_{\text{star}} - C_{\text{sky}})}{(C_{\text{comp}} - C_{\text{sky}})} \right]$$

Δm is the difference in magnitude between the comparison star and the astrovariable (remember that the dimmer object has the larger magnitude number), C_{star}, C_{comp}, and C_{sky} are the counts within the same-sized apertures around

Photometry

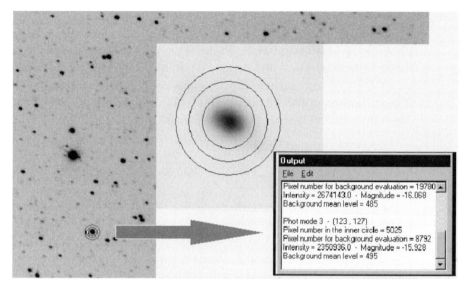

Figure 4.4 This is a screenshot of Christian Buil's IRIS aperture photometry program in action. The light from the star is measured within the inner circle. The sky background contribution is measured in the outer ring. The magnitude of the star can then be deduced by the software. This illustration is by courtesy of Christian Buil and is also featured in a paper in Vol. 113 No.2 of the *Journal of the BAA* (2003), authored by Nick James and W. J. Worraker, and reproduced here with their special permission.

the astrovariable, the comparison star, and the sample of empty sky, respectively. You should sample several areas of blank sky to check for consistency and take the average count as the true background reading. Why this relationship involves a logarithm is explained in Chapter 1.

Finally, knowing the magnitude of the comparison star and the value of Δm allows you to find the magnitude value of the astrovariable. You should repeat this procedure using at least one more comparison star. Better still use a third, or even a fourth comparison star. If any of the magnitude values you derive disagree keep going until you are sure where the error is. It is just within the realms of possibility that one of the assigned comparison stars is itself a variable! Pass any such discovery on to your observing group co-ordinator.

What size 'aperture' should you use?

Owing to the seeing and instrumental defects (including errors in tracking), star images will be larger than simple optical theory would predict. In fact each star image in the field of view will be a 'blob' whose size and shape can be described by the image *point-spread function*. Figure 4.5 shows a typical example. The ideal size of the aperture you select to surround each star image should be enough to encompass almost all of the light from this point-spread function but not so much as to dilute the count with an overlarge contribution of light from the sky background.

4.7 Obtaining magnitude measures from a CCD image

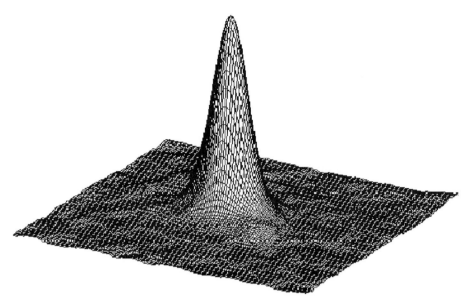

Figure 4.5 The point-spread function of a star's image. This represents how the intensity of the star's light varies across its image, height representing intensity. This illustration is by Nick James and W. J. Worraker and appeared in their paper in Vol.113 No.2 of the *Journal of the BAA* (2003) and is reproduced here with their special permission.

You will have to make the judgement by eye. Try momentarily turning up the brightness of your monitor to full (but *do not* use the image processing software to make any changes) and increase the size of the aperture until it looks to you that you have included all the light from the star within it. It is crucial that you use this same size of aperture to make each of the other counts. If a small amount of the light from the star does fall outside the aperture, then you can be fairly sure that this will also be the case for all the other stars you measure. Hopefully the ratios of brightnesses will not be skewed as long as you ensure that the aperture contains almost all of the light from each object.

A refinement to the sky background reading

It is best practice to apply *a median* function on the aperture you use to sample the sky background. The median count is a bit like an average count. In effect, though, a given cluster of pixels are all brought to the same level of brightness. The presence of any faint stars that might lurk in the aperture then has a lesser effect on the final figure than it would for a simple summation of brightness. It is highly likely that your software package will sport a median function for you to use.

Use only counts from 'clean' apertures

Even after our calibrations, some images will contain flaws. Believe it or not, even random cosmic ray hits can fire individual pixels or lines of pixels in the

CCD during an integration. If you see any such sources of pollution make sure that you do not place the sampling apertures to include counts from them. If any such blemishes are so awkwardly placed that you cannot avoid them, then it is best to discard the image and use another. This is another good reason for taking a series of images during your observing run. Failing that, the only remedy is to replace the affected pixels with the median level – in effect to paste out the blemishes. Such tinkering should really only be a last resort.

Consistency and calibration

Can you rely on the software to give you accurate magnitude values? I recommend testing this out on several fields containing comparison stars (there are hundreds for you to choose from on the CD-ROM that accompanies this book). In any field make one of the comparison stars your 'astrovariable'. Perform the observing run and reduction procedure using the other comparison stars as known standards. Does your chosen star come out at the correct magnitude?

Repeat this procedure on fields containing stars to cover a range of known brightnesses. It is worth spending several evenings doing just this before embarking on your astrovariable work. From your results, make a graphical plot of the true magnitudes of the stars against your own determinations of their magnitudes. Ideally all these magnitudes will perfectly agree. Even if they do not, the graph should show either a smooth curve or a straight line. If so, you can use this graph to calibrate your final results of astrovariable brightnesses and in this way produce a correct final value from the result of each observation.

Any scatter in the graphical points will give you an indication of the likely uncertainty in any subsequent astrovariable brightness measurement you do obtain. *Don't forget to include this uncertainty value in the report you send to your observing co-ordinator.* Your results are still worthwhile provided the uncertainties are no more than $\pm 0^m.1$. Potentially, though, they could be as good as $\pm 0^m.01$.

Up to the time I am writing these words, it is still true that some of the automated photometry programs manufacturers supply with astrocameras cannot be relied upon to give very accurate results. I cannot overemphasise the importance of investigating (and calibrating if necessary) your entire system in the field before you routinely begin submitting your observations to your co-ordinator. As a bonus you will get valuable practice while working with stars of known brightness before you embark on working with astrovariables.

4.8 Filters for photometry

Before the mid-nineteenth century the only detector available to astronomers was the human eye. Then the first photographic emulsions were invented and eventually improved to the point that they could, with long exposure times, record images of celestial bodies.

4.8 Filters for photometry

However, the earliest photographic emulsions were most sensitive to a wavelength of about 430 nm. The Earth's atmospheric transmission imposed a cut-off at about 350 nm (ultraviolet) and the emulsion sensitivity dwindled to zero at about 550 nm (yellow). This is very different to the spectral response of the human eye (which I described in Section 4.1). To distinguish brightness measurements made by eye and from photographs astronomers referred to *visual magnitudes*, m_V, (often known as 'V-band magnitudes') and *photographic magnitudes*, m_P, sometimes m_B (often known as 'blue magnitudes' or 'B-band magnitudes').

A further improvement in the precision of brightness measurements eventually resulted from the discovery of the *photoelectric effect* in 1888. When light falls on the surface of some substances (particularly the alkali metals) electrons are ejected. If a small disk of this substance is mounted inside a small glass vessel from which air has been excluded, then the liberated electrons can be collected by a positively charged electrode and the resultant current can be measured. The brighter the light falling on the photoelectric surface, the greater the current. By the 1940s a combination of photoelectric detector and amplifier, the *photomultiplier tube*, allowed accurate measurements of even the faint light levels received from stars.

Until just a few years ago the photomultiplier tube was the heart of the measuring device – the *photoelectric photometer* – used by professional astronomers and advanced amateur astronomers alike.

One problem was that photomultiplier tubes had a spectral response (response to light of differing wavelengths) that was different again to the old photographic emulsions and to the human eye. Astronomers had to devise filters with the correct bandpasses in order to make their photometers conform to the older standards of measurement. Hence was born the system of passbands defined by standardised filters and photomultiplier tubes originated by Johnson and Morgan and subsequently developed by others.

Now photoelectric photometers have been superseded by CCDs. It was back to the drawing board with regard to the filters we have to use with them in order that our measures conform to the old established standards. It is only very recently that matters have been settled. Since CCDs vary in their precise spectral responses from model to model (see Section 4.1), your best course of action is to consult the manufacturer's literature (or even consult the manufacturer directly if you need to) and obtain the recommended filters for your photometric work.

You will certainly want a filter set in order to carry out V-band photometry and maybe a set in order to carry out B-band photometry if you are interested in monitoring colour changes in some of your chosen astrovariables. However, do bear in mind that many CCDs have a rather poor B-band sensitivity – again consult the manufacturer as to what is appropriate for a given product. There are filters for other passbands you could obtain but my advice is to concentrate just on V-band photometry until you become well practised. Become a virtuoso of that before expanding your repertoire.

4.9 Just the beginning

Photometry is a very large and technically complex subject. I must make it clear that I have presented you here with just the specific techniques and items of information that you will need to get going in this field if you are a typical amateur astronomer. I have deliberately left out areas which you might find just of academic interest or those which will be of interest to you only when, or if, you get involved in more advanced work.

For instance, if you really want to know more about the practice of photoelectric photometry, then I refer you to Chapter 13 in my book *Advanced Amateur Astronomy*, published by Cambridge University Press in 1997. Similarly, I have left out more specialist techniques such as *point-spread function photometry* and *all-sky photometry*. Certainly the latter is most suitable for the professional astronomer in a world-class observing site, although the former technique can be of use to the amateur.

There is enough material in this chapter to give you a flying start in the practice of photometry. Work with care and precision and you will span the gulf that is usually asserted to separate amateur and professional astronomers.

Chapter 5
Stars great and small

The first examples of stellar variability were discovered centuries ago. To date we have identified and catalogued tens of thousands of variable stars, while millions more await detection and investigation. Variable stars show such diversity in their behaviours and the causes of their variability that just remembering the main characteristics of all the types is no easy task, let alone understanding them. In many cases the reasons for variability are still the subject of ongoing research, sometimes even contention, among researchers. In virtually all cases the fine details are still being worked out.

The complexities of the subject have resulted in a significant dearth of details and discussion about the reasons for variability in books written for the amateur astronomer. Literature aimed at the amateur variable star observer tends just to cover the basic mechanics of performing, recording, and submitting observations while ignoring the astronomy and astrophysics of the stars themselves.

For me it is the astronomy and astrophysics which provide the main fascination of this subject, so in this book I will do my best to provide a background of understanding, as well as covering the practical details. However, do be aware that in this little book I can only cover the rudiments of the astronomy and astrophysics involved. For the same reason I can only cover details about the main types of astrovariable. However, this should be enough to get you started on a lifetime's study of them.

We begin our studies with one star which, happily for us, is not significantly variable.

5.1 Our daytime star

Being so close to us as our Sun is, we have been able to study it in great detail. It provides a useful datum from which we can compare other stars and build up an

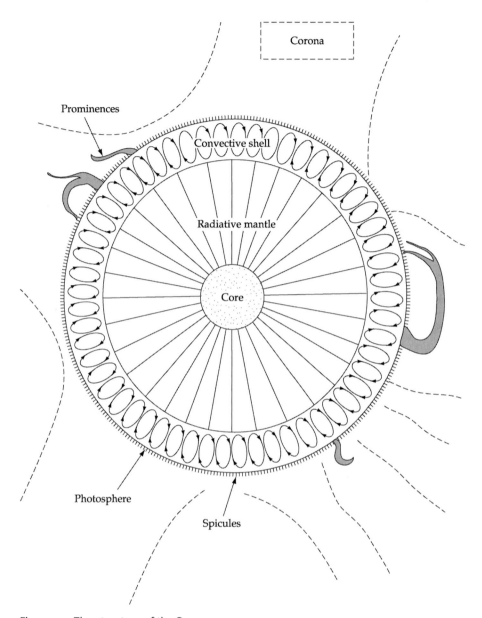

Figure 5.1 The structure of the Sun.

understanding of their physics and their evolution – and why they sometimes vary their luminosities.

The Sun's 2×10^{30} kg bulk of hydrogen (73 per cent by mass), helium (25 per cent) and other elements (referred to as 'metals' by astronomers, even though many of the elements are non-metallic!) occupy a vast spherical globe of radius 0.7 million kilometres. The values of mass and radius for the Sun define for it a mean density of 1.4 times that of water. However, the Sun is extremely

5.1 Our daytime star

Table 5.1 *Energy-producing reactions in the Sun*

Proton–proton cycle (PPI)
$$H + H \to D + e^+ + \nu$$
$$D + H \to {}^3He + \gamma$$
$$^3He + {}^3He \to {}^4He + 2H + \gamma$$

Secondary proton–proton cycle (PPII)
$$^3He + {}^4He \to {}^7Be + \gamma$$
$$^7Be + e^- \to {}^7Li + \nu$$
$$^7Li + H \to {}^8Be \to 2\,{}^4He$$

Tertiary proton–proton cycle (PPIII)
$$^3He + {}^4He \to {}^7Be + \gamma$$
$$^7Be + H \to {}^8H + \gamma$$
$$^8B \to {}^8Be + \nu \to 2\,{}^4He$$

CNO cycle
$$^{12}C + H \to {}^{13}N + \gamma$$
$$^{13}N \to {}^{13}C + e^+ + \nu$$
$$^{13}C + H \to {}^{14}N + \gamma$$
$$^{14}N + H \to {}^{15}O + \gamma$$
$$^{15}O \to {}^{15}N + e^+ + \nu$$
$$^{15}N + H \to {}^{12}C + {}^4He$$

Key:
H	Hydrogen nucleus
D	Deuterium (heavy hydrogen) nucleus
Li	Lithium nucleus
He	Helium nucleus
Be	Beryllium nucleus
B	Boron nucleus
C	Carbon nucleus
N	Nitrogen nucleus
O	Oxygen nucleus
e^-	electron
e^+	positron
ν	neutrino
γ	gamma-ray photon

Differing isotopes are indicated by the raised numbers in front of the species (eg. ^{4}He is the isotope of helium with an atomic mass number of 4)

rarefied in its outermost layers but its density increases enormously towards its core. The density ranges from about a millionth of a kilogram per cubic metre to about 150 000 kilograms per cubic metre.

Figure 5.1 illustrates the structure of our Sun, as we understand it from our theoretical models based on what we know about the behaviour of matter and what we can observe about our daytime star.

It is in the Sun's core that density of matter, temperature, and pressure are high enough to cause the particle collisions of sufficient ferocity that enable the nuclear transformations which release energy and cause our Sun to shine. The main energy producing reactions are detailed in Table 5.1. The cycles denoted PP stand for *proton–proton* because the reactants are hydrogen nuclei and these are single protons. The CNO cycle is often known as the *carbon cycle* because carbon is a catalyst, becoming various isotopes of carbon, nitrogen, and oxygen before reemerging at the end of the reaction chain as the original isotope of carbon.

If these reactions look complicated, they are. The end result is straightforward enough, though. It is the building up of one new atomic nucleus of helium from four atomic nuclei of hydrogen. However, the mass of a helium nucleus is very slightly less than the sum of the masses of the four hydrogen nuclei. This *mass deficit* is converted into energy in accordance with Einstein's famous $E = mc^2$. The conversion rate is handsome. For every kilogram of matter lost the energy released is 9×10^{16} J of energy.

The temperature in the core of our Sun is about 15 million K. We can say this is 15 million °C, since the Kelvin and Celsius scales share the same degree size, with the zero of the Kelvin scale being a mere 273 degrees below that of the Celsius scale. The density at the heart of the Sun is 1.5×10^5 kg m^{-3} (this is the same figure as before but expressed in scientific notation). These conditions are right for the PPI (proton–proton I) cycle to be the dominant energy producing reaction. PPI, PPII, and PPIII produce 70 per cent, 29 per cent, and 0.1 per cent of the Sun's power, respectively. The CNO cycle produces less than 0.1 per cent of the power in the case of the Sun. However, other stars have different conditions at their cores and in those cases the foregoing numbers will differ. In fact, for massive stars it is the CNO cycle which dominates.

5.2 Our stable Sun

We have observed very little in the way of changes in the size of the Sun and its brightness in the several centuries since we have been carefully observing it. Putting on our astrophysicists' hats, we can say that the Sun exists in a state of both *thermal equilibrium* and *hydrostatic equilibrium*. We will consider what hydrostatic equilibrium means shortly. First, though, let us take a closer look at matters thermal.

To say that the Sun exists in a state of thermal equilibrium means that all of the energy produced in the core of the Sun (and we think that little or no energy is produced outside the core) finds it way to the solar 'surface' (more properly called the solar *photosphere*) with nothing being 'used up' nor created by any processes in between.

As an aside, here I will distinguish between energy (measured in joules) and power (measured in watts). An energy consumption or production of 1 joule per second (1 J s^{-1}) is equal to a power of 1 watt (1 W). For instance 1000 J s^{-1} = 1000 W, more usefully expressed as 1 kilowatt (1 kW). A million watts is equal to 1000 kW, or 1 MW (1 megawatt).

5.2 Our stable Sun

It is a matter of practical physics to measure the energy per unit area we receive from the Sun each second (the *solar constant*). Its value is 1.368 kW m^{-2} (1.368 kilowatts per square metre). Using that figure and knowing the distance to the Sun (150 million kilometres) and the radius of the Sun (0.7 million kilometres), we can work out the power emitted per unit area of its surface. This is 6.3×10^4 kW m^{-2} (63 000 kW per square metre). Knowing the radius of the Sun, we can work out its total surface area and thus sum the total power output of our daytime star. It turns out to be a staggering 4×10^{20} MW.

To produce this power output the Sun is losing mass at the rate of 4.4 million tonnes per second. All of this power is produced in the Sun's core, much of it in the form of γ-ray and X-ray photons. Some energy is also carried away by those curious ghostly particles known as neutrinos. These pass straight through the bulk of the Sun and out into space, having little interaction with the solar material.

The γ-ray and X-ray photons undergo a lengthy and arduous *random walk* of interactions, absorptions, and reemissions in the body of the Sun until by the time they reach the solar photosphere they typically have much lower energies (though there are many more of them), mostly corresponding to wavelengths in the visible range of the electromagnetic spectrum.

Actually the emerging photons have a statistical distribution of the numbers of them occurring at each given wavelength. This distribution is almost the same as for a *black body* (a concept in physics, pertaining to a hypothetical perfect absorber and emitter of radiation) of temperature 5800 K.

This statistical distribution gives rise to a distribution of energy with wavelength the graphical representation of which is known as a *black-body curve*. This curve is dependent on temperature, as illustrated in Figure 5.2. The Sun and all normal stars produce black-body radiation like this. We call this smooth distribution of energy with wavelength the *continuum* emission of the Sun (or other star).

As illustrated in Figure 5.2, the wavelength of maximum energy emission shifts with temperature. The peak moves to shorter wavelengths as the temperature increases. A warm (room temperature) black body emits mainly in the infrared portion of the spectrum. A hotter (hundreds of degrees) body will emit enough shorter-wavelength radiation to begin to glow in the red part of the spectrum. As the temperature is further increased, so the body appears to brighten and become orange, then yellow. By this stage the temperature is some 5000 K. Increasing the temperature still further will make the black body appear white-hot, then greenish-white, then bluish-white and finally blue as the peak wavelength of emission shifts into the ultraviolet. The temperature would now be about 40 000 K.

Note, however, that the *peak* wavelength of emission is just that and not the *only* wavelength at which the radiation is emitted. This is why star colours are pastel shades and never appear vivid. The Sun's photospheric temperature of 5800 K results in the wavelength of peak emission being about 5.4×10^{-7} m (just over

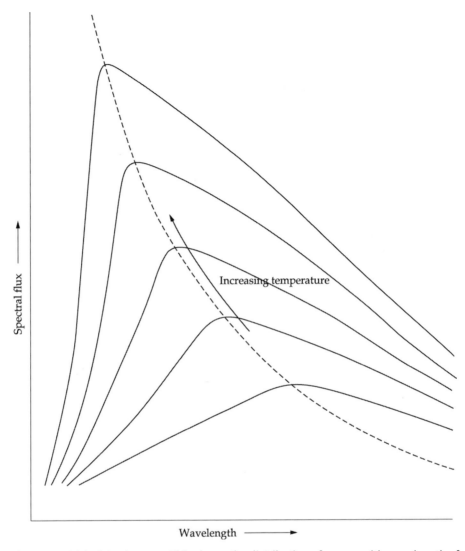

Figure 5.2 A black-body curve. This shows the distribution of energy with wavelength of radiation emitted by a black body at different temperatures.

half a millionth of a metre, or 540 nanometres) which is why it appears yellowish-white. This is also the wavelength to which our eyes are most sensitive, thanks to adaptive evolution.

The relationship between the sum total amount of energy emitted by a black body over all wavelengths and its temperature is given by the Stefan–Boltzmann law:

$$E = \sigma T^4$$

where E is the total energy emitted per square metre of the surface of the black body per second and T is its absolute (Kelvin) temperature. σ is known as

5.2 Our stable Sun

Stefan's constant and its value is 5.7×10^{-8} W m^{-2} K^{-4}. So, once we have found the surface temperature of a star (by observing its colour) we can calculate the amount of energy being radiated by each square metre of its surface.

There is some finer detail, known as *spectral lines*, impressed on the continuum emission of the Sun and other stars. I will have more to say about this later but we should now take a look at what we mean when we say the Sun is in hydrostatic equilibrium.

A star will remain physically stable if there is a balance between the pressures tending to make it collapse and those tending to make it expand. There is just one pressure that operates to make it contract: self-gravitation. There are two components that will make the star tend to expand: gas pressure and radiation pressure.

Any gas exerts a pressure by virtue of its temperature. The higher the temperature of any contained volume of gas, the higher the pressure it exerts on whatever is containing it. This is true even if the 'container' is the surrounding gas, as is the case inside bodies like the Sun. The photons of electromagnetic radiation colliding with the particles in the gas during their 'random walk' also exert a pressure. In the case of a star like the Sun it is gas pressure which dominates, though more massive stars have cores which produce very much more power and radiation pressure becomes more important.

Given that we know the Sun is stable we can assert that for any given thin layer of the solar interior we care to choose, the weight per unit area of the material above that layer bearing down on it is exactly balanced by the gas and radiation pressures exerted on the underside of that the layer. If this was not true then the Sun would not be stable. It would, instead, be a variable star.

Astrophysicists have five interrelated *equations of stellar state* with which they can coarsely describe the basic physics of stars. These are coupled differential equations. Unfortunately you would need to be at least up to university undergraduate level in physics and mathematics to make much headway with them. However, there are some fairly straightforward mathematical results that arise from these equations. Some of these are pertinent to a star's variability. More on this in the next chapter. I will content myself at this juncture by saying that there is a given profile of temperature with radius in the Sun and other stars that in a large part establishes their physical properties – including whether or not they are variable in brightness. This profile, and therefore the internal and external properties of the star, is chiefly determined by the star's mass.

As I stated earlier, in the case of the Sun the central pressure and temperature are such as to favour the PPI cycle of nuclear transformations. The energy-producing core extends to about a quarter of the Sun's radius.

Beyond that region the material of the Sun freely circulates in currents concentric with its rotation axis, though it undergoes little circulation in the radial direction. Any layer of material in this zone stays at that level. It does not move outwards towards the photosphere nor downwards towards the core. Here the main process transporting energy outwards from the core towards the photosphere is the photons of electromagnetic radiation undergoing their 'random walks'.

Incidentally, it can take the photons produced by the core from over a hundred thousand years to about a million years to jostle their way to the outermost zones of the Sun! Some of them take even longer. No wonder they have lost much of their energy by the time the do achieve their freedom! This region of the solar interior, occupying the main bulk of the Sun's body, is termed the *radiative zone*.

In the outermost 29 per cent of the Sun's radius the temperature is low enough to allow electrons to capture nuclei. So in this region the Sun is partly composed of ionised atoms, rather than the co-existing mixture of pure electrons and pure nuclei that is the case below this level. These ionised atoms are much more efficient at capturing photons and so this material is more opaque to the passage of radiation than the material in the lower levels. Consequently in the outermost 29 per cent of the radius of the Sun convection takes over as the main energy transport mechanism.

Here heated gas expands and loses density, rising up through the surrounding gas until it cools and is replaced by freshly heated gas, the cooled gas descending to the lower levels where it is heated once more. Of course, radiative transfer also happens in this *convective zone* but convection provides the way most of the energy is carried towards the 5800 K photosphere. Ultimately the photons leave the photosphere by being radiated into space, then to bathe us and all other bodies in the Solar System.

At the photosphere of the Sun, and in the overlaying *chromosphere*, it is chiefly the interactions between the hot ionised gas, known as *plasma*, and the Sun's convoluted and powerful magnetic field (which itself it affected by the motion of the churning plasma) which produces the phenomena of sunspots, faculae, plages, prominences, and solar flares. The Sun also possesses a tenuous but extremely hot *corona*, extending outwards into the Solar System and, in addition, sends out the gusty *solar wind* of particles that interacts with the planets and the planetary magnetospheres.

5.3 Spectral lines

A smooth rainbow of colours is known as a *continuous spectrum*. It is produced by any hot body that is made up of matter of relatively high density, such as is the case for a liquid or a solid. A smooth black-body distribution is an example of a continuous spectrum.

The situation changes when we consider the light produced by matter at low density – such as for gases at low pressure. These produce *line spectra*. In order to understand how they do this we must first consider the structure of an atom.

An atom of an element, say hydrogen, consists of a tiny positively charged nucleus surrounded by an electron (other elements have a greater positive charge on the nucleus and more electrons orbiting it). Although the electron can be thought of as a satellite moving round a planet, the real situation is not quite as

5.3 Spectral lines

simple as this. A better description of the electron is a cloud of charge shaped into a standing-wave pattern that surrounds the nucleus.

This standing wave can take a number of different forms depending on the energy of the electron. The electron can exist only in one of a number of discrete energy levels when it is trapped within an atom. Each energy level corresponds to a given standing wave. The electron can exist in any of these levels, depending upon how much energy it has.

It can accept energy to go from one energy level to a higher level. It can even give up energy to fall to a lower energy level, but it can never rest between energy levels. This is the reason for the production of discrete spectral lines. If we supply energy to the gas by heating it to a very high temperature, or maybe drive an electric current through it by applying a high voltage, then we can cause some of the electrons surrounding the atomic nuclei to be excited to higher energy levels. In becoming excited the electrons take up the energy being supplied. However, this situation is temporary. Soon afterwards the excited electrons give up their energy and fall back to lower energy levels. *This discarded energy is radiated away as photons of electromagnetic radiation.*

The energy of the emitted photon is equal to the energy liberated by the de-exciting electron. This is equal to the difference in the values of the energy levels through which the electron falls. The resultant wavelength of this photon depends on its energy and is given by Planck's law:

$$E = \frac{hc}{\lambda},$$

where E is the photon energy (measured in joules), λ is the wavelength of the photon (in metres), c = speed of light (3×10^8 m s^{-1}) and h = Planck's constant (6.63×10^{-34} J s).

Let me restate the main point: each of the atoms of a particular chemical element will only have certain precisely defined energy levels which the electrons in each atom could occupy. Therefore any electron can only undergo one of a number of well-defined energy transitions as it changes from one energy level to another.

In any given sample of the gas (which will contain millions, perhaps billions, of atoms) all the possible electron energy transitions will occur in a given small increment of time. In this way a characteristic series of wavelengths of radiation will be produced. If the light from the excited gas is viewed through a spectroscope, then a continuous spectrum will *not* be seen. Instead, images of the slit will be seen only in certain wavelengths. A series of coloured lines will be produced, each in the position corresponding to its colour and wavelength.

A spectrum of hydrogen consists of four visible lines: one in the red, one in the blue, one in the blue-violet, and one in the violet. Various other lines occur in parts of the spectrum to which our eyes are not sensitive. Also, some electron energy transitions are preferred over others. This leads to spectral lines having differing intensities. The red line in the spectrum of hydrogen, known as Hα, is much more intense than the others for this reason.

Other elements have different arrangements of electrons in their atoms, and so each element produces its own characteristic spectrum. This affords us a wonderfully powerful way of finding out what chemical elements are present in any remote source (such as the stars). If the lines of a certain chemical element show up in the spectrum of a source, then that element is present in that source.

When a spectrum is seen as a series of bright lines against a dark background it is known as an *emission spectrum*. It is also possible to get the appearance of a series of dark lines on a bright continuous spectrum. This is called an *absorption spectrum*. Most stellar spectra are of this type, so it is as well to consider how the situation arises.

Just as light may be given off when an electron becomes de-excited, so the incidence of light of exactly the right wavelength on an atom may cause an electron to jump to the corresponding higher energy level. If a little salt is sprinkled into a flame, then the flame is coloured golden yellow by sodium ions being released into it and becoming excited. If this flame is viewed in a spectroscope, then the characteristic spectrum of sodium is seen, consisting of two bright lines very close together in the yellow part of the spectrum.

If the flame is exchanged for a powerful filament lamp, then a continuous spectrum is seen. This is just what we would expect to see produced by the hot metal filament. However, if the salt-sprinkled flame is then placed between the lamp and the spectroscope slit, so that the lamp is shining through the flame, something different again is seen. The continuous spectrum is seen to be crossed by two dark lines in the yellow region – exactly where the bright emission lines were seen when the flame alone was viewed!

What is happening is that some of the light from the lamp, at just the right wavelengths, causes more energy jumps in the electrons surrounding the sodium nuclei. As a consequence this light is absorbed. When the electrons subsequently de-excite, the light that was sent by the lamp in the direction of the spectroscope is now reradiated in *all* directions. Hence the light at those specific wavelengths is depleted when it reaches the spectroscope. In this way dark absorption lines are seen on the bright continuum.

The situation with a star is that the base of the photosphere acts, like the filament lamp, as a source of a bright continuum. I say 'base of the photosphere' because the photosphere is actually a thin layer. In the case of our Sun it is just under 200 km deep. The base of the photosphere of our Sun exists at a temperature of about 6400 K, while the top of it is at a temperature of about 4400 K. The lowest part is more opaque than the upper part. At an intermediate level where we see most of the light the temperature averages about 5800 K. The upper, cooler, layers of the photosphere and the overlaying chromosphere of the Sun, or other star, act like the flame in our example and this is responsible for impressing an absorption spectrum onto the continuum (see Figure 5.3).

5.3 Spectral lines

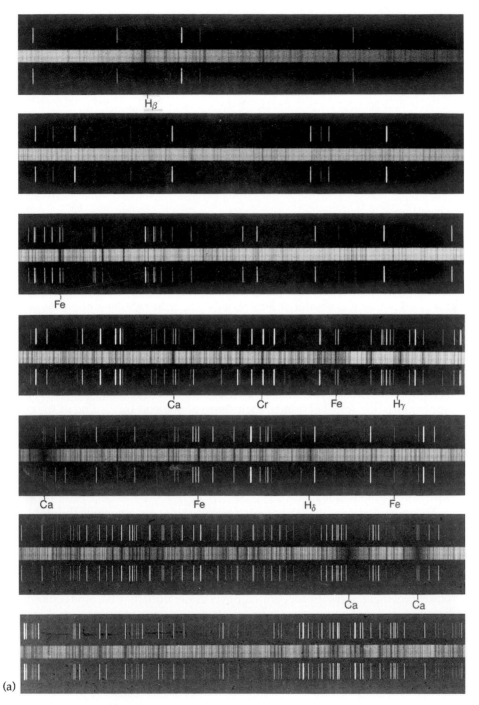

Figure 5.3 (a) High resolution spectrum of sunlight reflected from the surface of the Moon, taken by the author, using the 30 inch (0.76 m) coudé reflector and high dispersion spectrograph of the Royal Greenwich Observatory. The wavelength range covered is approximately 355 nm (left-hand side of the bottom strip) to 504 nm (right-hand side of

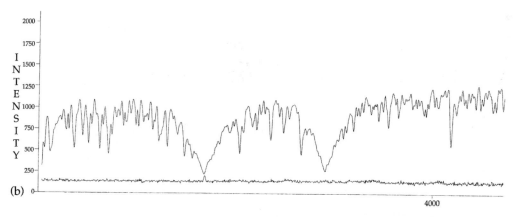

Figure 5.3 (cont.) the top strip), the wavelength increasing from left to right along each strip. There is a small amount of overlap between each strip. Above and below the main spectrum is a copper–argon emission spectrum exposed on the plate at the same time as the Moon spectrum for calibration purposes. (b) A small part of a tracing from the spectrum. The two broad dips in the spectrum of sunlight (upper trace) correspond with the broad calcium (Ca) absorption features indicated at the right-hand end of the second strip from the bottom in (a).

The single atoms that we have been considering so far form *line spectra*. Molecules (two or more atoms bonded together) are capable of more complicated transitions and they tend to produce *band spectra*. Band spectra have the appearance of a series of bands, each sharp at one end but fading at the other. Closer examination reveals that the bands are in fact made up of numerous fine lines very close to one another. Like line spectra, band spectra can be seen in either emission or absorption. However, most molecules are broken down into their component atoms before any really high temperature is reached, so only the coolest stars display band spectra.

The foregoing describes the case for matter at low density when the atoms are on average far apart. When atoms are pushed closer together, their interactions produce a merging, or 'blurring', of their electron energy levels. Consequently electrons can then have a range of allowable energy levels. This produces a progressive *pressure broadening* of the spectral lines. When the atoms are much closer together the energy levels merge ever closer until they become continuous. This is the reason why matter at higher densities produces continuous spectra.

5.4 Stellar spectra

Stars come in a wide range of masses, sizes, and surface temperatures. Most stars, though, have chemical compositions which are rather similar to that of the Sun. Yet, they can exhibit very different spectra. The main reason for this is that different temperatures favour different electron energy level transitions and so given sets of spectral lines each become apparent over a relatively limited range of photospheric temperatures.

5.4 Stellar spectra

Table 5.2 *Stellar spectral types*

Spectral type	Typical photospheric temperature (K)	Colour	Spectral characteristics
W	50 000	Blue	Wolf–Rayet; bright emission lines
O	50 000	Blue	Helium lines prominent, together with carbon, silicon, and highly ionised ions of lighter elements
B	23 000	Blue-white	Helium fades and hydrogen becomes more prominent; oxygen and nitrogen replace the more highly ionised species
A	11 000	White	Hydrogen dominant; lines of calcium, iron, chromium make their appearance
F	7 600	Yellow-white	Weaker hydrogen; calcium lines become very strong; very strong lines of neutral and ionized metals and heavy atoms
G	6 000	Yellow	Metal lines prominent; CN and CH bands make an appearance
K	4 000	Orange	Metal lines dominant; some molecular bands also very strong; titanium oxide bands appear
M	3 500	Orange-red	Titanium oxide bands dominant; many other lines of neutral metals and bands of molecules are prominent
L	1 700	Red	Complex spectra, including metal hydrides and metals
T	1000	Deep red	Complex spectra including water

There is an additional classification – the C stars. These are normally the coolest stars of spectral type M and so may be considered to be a subtype of M stars. They have abnormally strong spectral features due to carbon. The W stars are also a special case.

Historically, stellar spectra were classified by letters in alphabetical order. The sequence has several times been rearranged and amended in the light of new knowledge and interpretations. Today the spectral sequence has become:

O, B, A, F, G, K, M, L, T

Table 5.2 provides a brief description of the characteristics of each spectral type.

There is a spectral type W, which is now regarded as a special case. This corresponds to *Wolf–Rayet stars*. There are also types R, N, and S (though the use of these letters is now going out of fashion – they are nowadays most often collected together as type C) which can be thought of as branching off spectral type M.

T-type stars and the cooler examples of L-type stars are notable because, unlike the Sun and other normal stars, they do not shine as a result of normal fusion reactions in their cores. The hottest of them produce energy by sedately fusing deuterium nuclei, though this produces much less energy than normal hydrogen fusion. The rest shine only as a result of their thermal reserve of energy resulting from their initial formation and their continuing contraction. These not-quite-stars are called *brown dwarfs*.

There is a progressive change in the appearances of the spectra going along the sequence from O to T. The basic types have each been further divided into ten subgroups. Thus type A0 is very similar to type B9 and A5 is halfway between types A0 and F0.

In M-type stars the chromospheric temperatures are low enough to allow some very stable molecules, such as zirconium oxide, to form. So in the spectra of these stars we find the characteristic bands that are the signature of these molecules. These molecules cannot form in hotter stars, even though zirconium and oxygen will be present in very similar proportions in the hotter stars.

C-type stars are aged stars with sooty envelopes (a consequence of changes with age). That is why they can be thought of as a spur off the type M grouping of stars. Their spectra show the presence of carbon, as well as other stable molecules such as CH and CN in addition to the other features occurring in M-type stellar spectra.

To take just one example, it is the higher photospheric/chromospheric temperature of the A0 star which causes it to display the *Balmer* lines of hydrogen so prominently, while inhibiting the formation of the lines due to other elements that are more evident in the solar spectrum. Yet the basic compositions of the Sun and Vega are very similar, though you would not think so by comparing their spectra (Figure 5.4).

5.5 Information from spectra

A spectrum is considered to be of good quality if it is *pure*. This means that the light registered at any position on the spectrum consists of light of one specific wavelength, with no contamination by light of other wavelengths. How 'spread out' the separated wavelengths are is properly known as the *dispersion* of the spectrum.

As already discussed, the spectrum of a source gives information on its composition and temperature (defining the 'source' to be the light-emitting object under study). Given a good quality spectrum of sufficiently high dispersion, there are measurements which can be made in order to quantify many of the physical conditions existing in the source. These measurements are usually made on a graphical representation of the spectrum (see Figure 5.3(b)).

In the case of a star, measuring the intensities (area of the feature existing below the continuum level on the graph, in the case of an absorption line) of the various spectral lines present reveals how much of each of the various elements is

5.5 Information from spectra

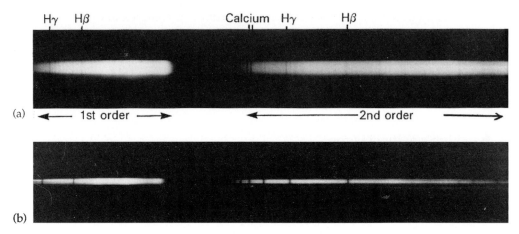

Figure 5.4 Comparison between the spectra of (a) the Sun (a G2 star) and (b) Vega (an A0 star). Notice that Vega's spectrum is dominated by the Balmer lines of hydrogen but many of the lines due to other elements in the solar spectrum are absent in Vega's spectrum. Spectra obtained by the author using the spectrograph on his 0.46 m reflecting telescope.

present (checked against the temperature as determined by measuring the star's colour and/or continuum spectrum because line intensities are also temperature dependent).

When seen at sufficiently high dispersion, spectral lines show some shape and structure. This is noticeable in Figure 5.3(b). The way the intensity varies with wavelength is known as the *line profile* of the particular spectral line. This is due to the presence of *hyperfine structure*, itself related to quantum effects within the emitting atoms. The physics of this is beyond the scope of this book but the presence of such hyperfine structure is useful in that it allows us to test the presence of, and maybe even to measure, magnetic fields. The hyperfine structure is further spread out by an amount dependent on the strength of a magnetic field present at the source. This is the *Zeeman effect*.

Pressure broadening also provides information about the pressure existing in the emitting region. Although both pressure and the presence of magnetic fields cause spectral lines to broaden, the two effects can be distinguished by the careful measurement of a high-quality spectrum. The way spectral lines are *polarised* also adds some detail about the physical conditions at the source – but again this is taking us into deeper physics than we need to indulge in at present.

You are probably already well aware of the *Doppler effect* and how this leads to *redshift* or *blueshift* in the measured wavelengths of spectral lines (compared with the same lines from a stationary source) if the source is receding or approaching, respectively. Careful measurements of high quality stellar spectra can do more than tell us the radial velocity of the star as a whole. It can reveal the motions of matter falling into and/or being expelled from the star. This is very useful in professional studies of variables stars and is particularly relevant in the cases of cataclysmic variables. Much more about those later in this book.

Although the interpretation of a spectrum is a complex and often difficult task, the clues about the physical conditions prevailing at a star or other source that it provides are invaluable to astronomers. That is why I have spent just a little space giving some superficial details of spectroscopy, even though it is not a field we will undertake for our main practical work of observing variable stars. It is, however, something advanced amateur astronomers could undertake.

5.6 Luminosity classes

There are a few stars which are the better part of a million times as bright as our Sun. Yet we should not regard our Sun as being puny. Some stars have been detected with only a millionth of the Sun's luminosity.

A star of any particular spectral type might have any one of several values of luminosity. Consequently, astronomers have divided stars into *luminosity classes*.

Remember that the spectral sequence is, for the most part, a temperature sequence. Also, remember that Stefan's law relates the amount of energy emitted per unit area of a star's surface to the fourth power of its absolute (Kelvin) temperature. If astronomers determine a star's spectral type then they are also determining its photospheric temperature. Put another way, two stars of the same spectral type must share the same photospheric temperature. Therefore if two stars of the same spectral type differ in brightness it must be because they have different surface areas. Further, the brighter star must have a larger surface area than the dimmer one.

If one star is 100 times brighter than another of the same spectral class, then it must have 100 times the surface (photospheric) area. Since the surface area of a sphere is proportional to the square of its radius, this means that the brighter star has 10 times the radius of the dimmer one. In this way luminosity classes relate to the physical sizes of the stars.

Class I stars These are *supergiants* with radii that are typically between 20 and 1000 times as large as our Sun. Rigel, of spectral type B8, is a blue supergiant of about 20 solar radii. For historical reasons stars at the blue end of the spectral sequence are said to be of *early* spectral type and those at the red end are said to be *late*. In this luminosity class stellar radii increase, going from early to late spectral type. Betelgeuse (α Orionis) is an M2 supergiant with a radius of nearly 500 times that of the Sun, though its lower photospheric temperature makes it only about a fifth of the brightness of Rigel. However, this is still 10 000 times as bright as the Sun!

Class II stars These are *bright giants*. They mostly range between about 1000 and 10 000 times the luminosity of the Sun and have radii between 10 and 100 times that of the Sun (but the coolest examples have radii up to about 400 times that of the Sun). ε Canis Majoris is a blue example of this type of star. There are fewer examples of stars of this luminosity class than of the others.

5.6 Luminosity classes

Class III stars These are the *giants*. They are mainly confined to spectral types G, K, and M. In this type of star, both the luminosity and the radius increase going from type G to M. Typical values are 30–1000 times the solar luminosity and 5–50 times the solar radius. Though most of the giants are yellow or red stars, blue examples do exist. γ Orionis (spectral type B2) is a case in point. Arcturus is more typical, being of spectral type K2. The blue giants are all much brighter than their red counterparts, approaching 4000 times the solar luminosity in the case of γ Orionis.

Class IV stars These are known as *subgiants*. This is another thinly populated class, with typical luminosities ranging between 5 and 5000 times that of the Sun. Their luminosities increase from type K to type B, though their radii remain around 5 times that of the Sun for all spectral types.

Class V stars These form the major class and are consequently known as *main sequence stars* (also known as *dwarf stars*). Most stars are of this class. Spectral type O main sequence stars have luminosities of about 300 000 times that of the Sun and radii about 20 times bigger. The radius and brightness of these stars decrease with later spectral types. A type M main sequence star has a brightness of the order of about 1 per cent that of the Sun and a radius about one-third of the solar radius. The Sun is a typical main sequence star, of spectral type G2.

Class VI stars These form a small group whose members are slightly smaller and less bright than the main sequence stars. They are known as *subdwarfs*. Most of them are of spectral types later than F.

Class VII stars These form the final classification. They are the *white dwarfs*, stars of very high density and tiny radius, typically one-hundredth that of the Sun. They have special significance for our ideas of stellar evolution, and there is more about them later in this book. They are all rather dim objects, usually having less than 1% of the Sun's luminosity.

Now that the luminosity classes have been covered in superficial detail, I can rectify an earlier oversimplification that I made in order to expedite the explanation of how astronomers deduce the sizes of stars from their intrinsic brightnesses and their spectral types. When I stated that the stars of a given spectral type are all at the same temperature I was being deliberately inaccurate in order to make the point clearly. In fact, the lower photospheric pressures in the larger stars actually cause them to mimic stars of a slightly different type.

As an example, an M0 supergiant star's photosphere is at a temperature of 3300 K. An M0 giant star has a photospheric temperature of 3600 K, while a main sequence star of the same spectral type has a photospheric temperature of 3900 K. Fortunately, there are enough clues in the spectra to distinguish between the luminosity classes and so astronomers can make the appropriate corrections when calculating the temperatures and brightnesses of stars.

Stars great and small

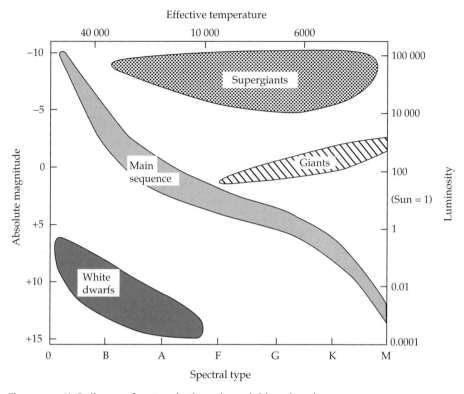

Figure 5.5 H–R diagram for stars in the solar neighbourhood.

The luminosity class of a particular star is often given with its spectral type. As an example, the Sun is $G2_V$.

5.7 The Hertzsprung–Russell diagram

A graphical plot of stellar luminosities against spectral types (which, you will remember, are closely related to photospheric temperatures) for any given collection of stars shows something very interesting. If there was no relationship between luminosity and spectral type, then such a plot would be formless; just a mass of dots covering the graph with each dot representing a particular star. However, this is not what is found in practice. In 1911 the Danish astronomer Ejnar Hertzsprung plotted the spectral types and luminosities of stars in several star clusters. He chose to plot the stars of clusters because he could be sure that the measured apparent luminosities of the stars in each cluster were all in direct proportion to their absolute luminosities because all the stars in a given cluster are at almost the same distance from us. He found a non-random distribution. The American astronomer Henry Norris Russell independently found the same result in 1913.

Today a plot of luminosity (or the equivalent absolute magnitude) versus spectral type (or the equivalent photospheric temperature) for a collection of

5.7 The H–R diagram

stars is known as a *Hertzsprung–Russell (H–R) diagram*. An alternative name for it is a *colour–magnitude diagram*, though I prefer H–R diagram and will use this moniker from here on.

An H–R diagram plotted for the stars is our neighbourhood of the Milky Way is shown in Figure 5.5. Notice how the stars are collected into groups on the diagram. Stars can exist outside these groupings and, as we shall see, stars can move across the diagram, sometimes even crossing from one group to another. However, the groupings are significant because *most* of the stars preferentially occupy these groupings *most* of the time.

The insights that H–R diagrams give us about the nature and evolution of stars, and even clues about their structures, are numerous and multi-faceted. It turns out that certain types of variable stars occur in particular locations on the H–R diagram – and the paths of many stars cross these regions of variability as they naturally evolve.

So, it is with a look at how some stars evolve into variables that, in the next chapter, we begin our task of understanding and classifying them. Along our way we will also meet our first candidates for practical observation.

Chapter 6
Variable beginnings

I hope that in Chapters 1, 2, 3, and 4 I have been able to help you arm yourself with the equipment, resources, and techniques to begin observing astrovariables. Here, building on the foundations laid in Chapter 5, we develop the story of why some single stars are variables and meet our first candidates for observation.

6.1 Single-star variables on the H–R diagram

When variable stars are plotted on an H–R diagram, they are seen to occur in certain locations on it. Figure 6.1 shows some of the main groupings. The variable stars in each grouping display similar characteristics of brightness variability. As I explained in Chapter 1, variable stars are most often classified into groups with a particular named star typifying the group. Group members also show similar macroscopic characteristics (mass, luminosity, and spectral type) by virtue of their common position on the diagram.

Stars move about on the H–R diagram as they evolve because their macroscopic characteristics change with time. Consequently even normally steady stars can undergo episodes in their lives when they cross into one or more of the regions of stellar instability and so become variables. Even our now-steady Sun has been a variable in its past and it will be again in the distant future. In order to make some sense of this it will benefit us to take a brief look at how stars are born and evolve.

6.2 Stellar nurseries within the interstellar medium

In the region of our Sun, the stars are separated from each other with an average distance of around 5 light years. This space is not empty but is filled with extremely tenuous gas and dust. Moreover, this gas and dust is not evenly distributed but is very 'lumpy'.

6.2 Stellar nurseries within the interstellar medium

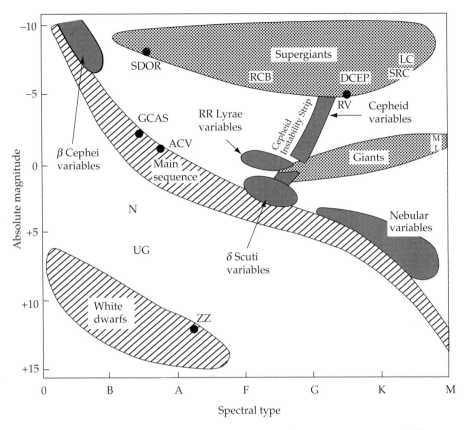

Figure 6.1 Some types of single-star variables shown in their positions on an H–R diagram.

Radio observations made at a wavelength of 21 cm reveal vast clouds of hydrogen gas with masses ranging from one tenth the mass of our Sun to over 1000 times the solar mass. The atoms in these clouds are typically at temperatures of around 80 K and the density of the gas is around 50 million particles per cubic metre. This may sound dense, but 50 million hydrogen atoms have a mass of only 8×10^{-17} kg.

There are also more massive clouds of molecules, mainly hydrogen, H_2, though with traces of more complex species. These clouds are very cold. Temperatures below 10 K are typical. These molecular clouds tend to be very big, often having masses the better part of a million times as large as the Sun. Dozens of different molecular species have been identified, most of them compounds of hydrogen, carbon, oxygen, and nitrogen.

Between them the atomic hydrogen clouds and the molecular clouds do not occupy all the space between the stars. The remaining space is filled up with a low-density (around 3000 particles per cubic metre) gas with a temperature of the order of 1 million K. This discovery arose from ultraviolet and X-ray studies made by using artificial satellites. This high-temperature gas permeates the entire Galaxy. Observations reveal that this gas is not steady, but rather seethes with turbulent motion. The free electrons in this turbulent plasma (neither molecules nor atoms can survive intact at these temperatures) cause the radio signals from distant sources to scintillate in the same way that the Earth's atmosphere causes the stars to appear to twinkle.

Most of the interstellar medium is gaseous, though about 1–2 per cent of it is in the form of solid 'dust' particles. As far as we can tell from their effects on visible light and their emissions in the infrared, the particles are mainly metallic silicates (aluminium, magnesium, etc.) and carbon (probably in the form of graphite). They also appear to have various gases frozen onto them. The particles are very small. They range in size from a thousandth of a micrometre to a few tenths of a micrometre.

Astronomers estimate that about half of the light from the stars in our Galaxy is absorbed by interstellar dust particles. In the process, the particles are heated. They reach an equilibrium at a temperature of about 30 K, when they then emit radiation in the far infrared part of the spectrum. At this temperature the particles are emitting energy at the same rate that they are absorbing it. This situation is called *thermal equilibrium*.

The interstellar dust does not absorb all wavelengths of radiation equally well. The short wavelengths are absorbed more than the longer wavelengths. This causes the light from distant stars to be both dimmed and apparently reddened.

The denser parts of the interstellar medium are readily observable from Earth. Perhaps the most famous example is the Great Nebula of Orion (Messier 42). Figure 6.2 shows a wide-angle photograph of the region around M42, taken using a telephoto lens. It lies about 1600 light years from us and the parts of it that are bright enough to show up in this photograph span about 26 light years across. As you are aware, there are plenty more such nebula in the sky.

Very-long-exposure photographs of Orion show that M42 is just one bright concentration in a large and complex nebula that spans the whole constellation. Infrared images show the cloud well and reveal several 'bright spots' not shown in visible wavelengths. These are almost certainly sites of star formation, as are other dense molecular clouds.

The general scenario that leads to star formation, as far as astronomers understand it, is that large clouds fragment into many self-gravitating cloudlets due to turbulence and instabilities. Each of these further contracts into spinning disks, thicker in the middle and thinner towards the edge. Gravity helps to condense ever more material towards the centre of the disk. Eventually a massive, self-gravitating, body forms at the centre. This is the growing embryo that will become a new star.

More than one set of initial conditions may spark the formation of stars. Stars are formed sometimes in massive molecular cloud complexes and sometimes in smaller units. Many of the small, dark nebulae we call *Bok globules* are known to be contracting and will ultimately give rise to stellar birth. Look closely at

6.2 Stellar nurseries within the interstellar medium

Figure 6.2 A wide-angle photograph of the sky in the region of M42, taken by the author using a telephoto lens.

Figure 6.3, which shows a small part of the Eta Carinae Nebula and you will see a few Bok globules close to the main part of the dark cloud. Perhaps some of these are contracting towards giving birth to new stars.

Stars do not tend to form singly. A whole group of them tends to form at more or less the same time. These give rise to the formation of *open star clusters*, groups of maybe a few dozen to several hundred stars in the same small region of the sky. One magnificent example of an open star cluster is that in the constellation of Taurus, the Pleiades. This cluster is 410 light years from us and is over 13 light years across. Figure 6.4 shows a photograph of the region obtained using a telephoto lens. Look carefully and you will see traces of nebulosity amid the cluster. It used to be thought that this is the material left over from the initial cloud that created the stars, roughly 20 million years ago. However, a study now indicates that the original nebulosity has been replaced by fresh material wafting through the cluster.

Star clusters are not permanent constructions. Over a period of time the proper motions of the individual stars cause star clusters to disperse.

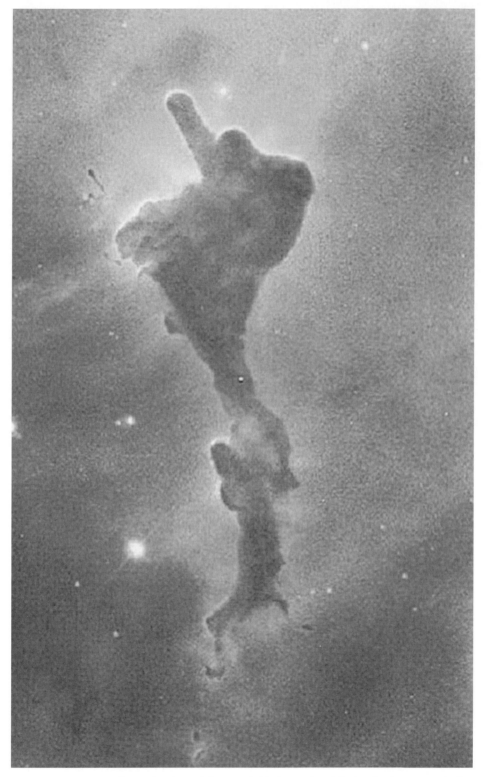

Figure 6.3 The author has nicknamed this formation 'The Rude Gesture Nebula'. It lies within NGC 3372, the Eta Carinae Nebula, otherwise known as the 'Keyhole Nebula'. Note the Bok globules just to the left of the main pillar of dark nebulosity. This Hubble Space Telescope image is courtesy of NASA and the Hubble Heritage Team (Aura/STScI).

Figure 6.4 The Pleiades star cluster in Taurus, photographed by the author using a telephoto lens.

6.3 An unstable start in life

An embryonic star can be said to exist as a separate entity when a core begins to develop in its contracting cloudlet. As material piles down onto this core so its temperature increases. When the core temperature reaches about 7 million K the first significant nuclear reactions begin and energy is produced via the proton–proton cycle. The star will continue to grow provided there is enough matter available in its mother protostellar cloudlet. Stars with masses less than about 75 times the mass of the planet Jupiter (this is about 0.07 times the mass of our Sun) can never achieve core temperatures high enough to initiate even the proton–proton cycle. These are the brown dwarfs (of spectral types L and T) I mentioned in the last chapter. You might call these 'not-quite-stars'.

If a star builds to a larger mass it can begin to shine as a true star. On the H–R diagram it will move from the far right and head towards taking its place on the main sequence. Even before it gets to this stage the star will be shining, its outpouring of heat and light provided by the gravitational energy liberated by the infalling material that is building it and by its own contraction.

It is at this point that we meet our first types of variable stars. They are all classed as eruptive variables because of their behaviour. These ones are also known as *nebular variable stars*, or sometimes *Orion variable stars*. In the GCVS classification Orion variables are indicated by the letters IN with other letters added to indicate the subtype. They are subdivided into INA, INB, INT, and IN(YY) because of certain characteristics, particularly in their spectra. INA stars are of spectral types O–A, while INB stars are of spectral types K–M. INT stars are the longest-recognised class of nebular variables, the *T-Tauri* variables, so named after their famous exemplar.

Nebular variables are physically associated with the nebulosity from which they are condensing, or have condensed. The IN(YY) stars have spectra which show tell-tale characteristics of nebular matter falling down towards the star's photospheres.

There are many examples of nebular variable stars known but most do not make easy quarry for the variable star observer. Some are immersed in bright nebulosity which makes visual or even photometric study difficult. As a class, they also tend to be faint and have rather small amplitudes of brightness variation, again precluding visual study (though not precision photometry). However, the CD-ROM that accompanies this book does contain the finder charts for a couple of less difficult examples: V586 Ori and V351 Ori. You will find them in the TA chart folder.

As an aside, please do remember that you can use the 'find' facility in your web-browser to quickly and easily locate specific items on the CD-ROM.

V586 Ori is of type INA, while V351 is of type INSA. The 'S' is another subdivision of the class, meaning that the star displays very rapid brightness changes.

Figure 6.5 (a) and (b) shows long-term and short-term light-curves of AB Aurigae, another example of a nebular variable star. This one is of type INA. Notice the episodes of fading. Such behaviour is characteristic of nebular variable stars. There is another INSA-type star's light-curve, this time the one for RR Tauri, also on the CD-ROM.

FU Orionis stars are a related class and are probably in reality just one (the earliest?) evolutionary stage of nebular variables. They tend to have A–G-type spectra and brighten by several magnitudes over a few months and then very slowly decline over months or years, though sometimes remaining at maximum brightness for months or years. There aren't many examples known at present but the star V1057 Cyg is one you might find interesting to observe. Its light-curve is shown in Figure 6.6 and you will find a TA finder chart for it on the CD-ROM that accompanies this book.

After the core has attained the threshold temperature of 7 million K nuclear reactions begin. The power output of the core increases rapidly with temperature, the star properly coming to life when the core reaches about 10 million K. In its birth pangs the star strives to reach an equilibrium with the force of gravity (tending to make it contract) and gas pressure (wanting to make it expand). At first the star is very unstable and its luminosity erratically varies. This brings us

6.3 An unstable start in life

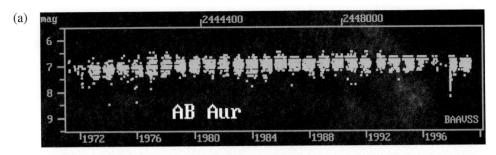

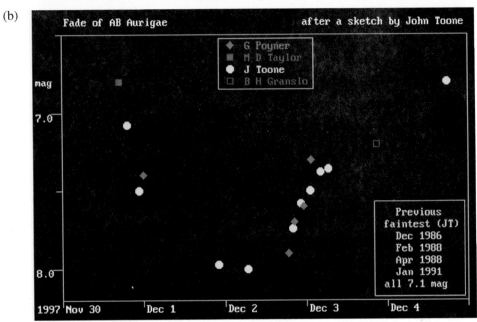

Figure 6.5 The light-curve of AB Aurigae; (a) spanning 28 years; (b) covering one fading event. Courtesy BAAVSS.

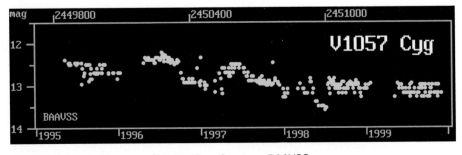

Figure 6.6 The light-curve of V1057 Cyg. Courtesy BAAVSS.

Variable beginnings

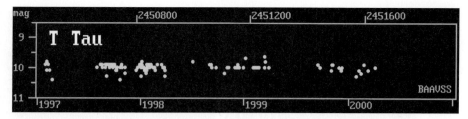

Figure 6.7 Light-curve of T-Tauri. Courtesy BAAVSS.

to the INT, or T-Tauri variable stars, probably the latest evolutionary stage of the so-called nebular variable class of pre-main sequence stars.

The behaviour of T Tauri, itself, between the years 1997 and early 2000 is shown by the light-curve in Figure 6.7. What the light-curve does not show is that this star also undergoes episodes of flickering brightness. The amplitude can erratically vary by several tenths of a magnitude at these times, changing as fast as $0^m.2$ in under a minute! Many nebular variable stars show such flickering and if you decide to take them on in your observing itinerary it is as well to make several brightness measures per night to check for such episodes. If you find the star is flickering it would make a fascinating evening's programme to stay with it and monitor it while this stellar baby is hiccuping.

Historically T-Tauri has remained between magnitudes $9^m.3$ and $13^m.5$. There is no finder/sequence chart for it on the CD-ROM but you can locate the star on charts or planetarium programs that go to at least tenth magnitude. It is located at a right ascension of $04^h\ 16^m$ and a declination of $+19°$. A very convenient photograph showing T-Tauri's position relative to the nearby Hyades star cluster, with the star identified, is given on page 21 of the August 2003 issue of *Sky & Telescope* magazine. Your observing group co-ordinator will probably be able to provide you with a sequence chart for this star if you decide to add it to your itinerary. You will find the light-curve of another INT star, VY Tauri, on the accompanying CD-ROM.

As well as hiccuping, newly born stars tend to emit powerful stellar winds. This often shows in their spectra as Doppler-shifted spectral features created by the emitted gas. In a few instances the evidence is even more direct. Such is the case for the nebula N81 in the Small Magellanic Cloud. As you can see from Figure 6.8, the stellar winds are clearing out the very obvious hollow within the nebula. The short-wave radiations from the stars also help us to see what is going on by causing the gas in the nebula to fluoresce.

6.4 Stellar adolescence and the ZAMS

As the core of the star comes to life so a combination of radiation pressure and hydrostatic pressure causes first the core, and then the star as a whole, to eventually cease its self-gravitating contraction. The star settles down to a sedate existence in the main sequence of the H–R diagram. More precisely, the

6.4 Stellar adolescence and the ZAMS

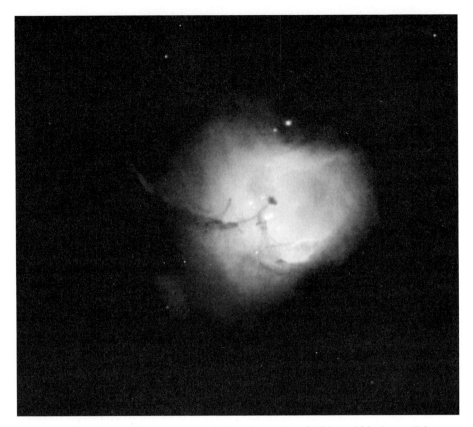

Figure 6.8 The Nebula N81 in the Small Magellanic Cloud. This Hubble Space Telescope image is courtesy of NASA and the Hubble Heritage Team (Aura/STScI).

star begins the main part of its life by joining the part of the main sequence known as the *zero-age main sequence* (*ZAMS*). This is a line following the lowest border of the main sequence band.

I mentioned in the last chapter that the equations of stellar state for a stable star are a complicated matter – and so they are. However, some of the results that can be derived from these equation are in themselves reasonably straightforward. For instance, the core temperatures of stars can be predicted for all stable stars made of normal matter (not white dwarfs or neutron stars), just knowing the mass and radius of the star. The relationship is:

$$T \propto \frac{M}{R}$$

where M is the mass of the star in solar masses (i.e. the mass of the Sun $= 1$), and R is the radius of the star in solar radii (the radius of the Sun $= 1$). T is then the absolute (Kelvin) temperature of the core of the star relative to that of the Sun. The Sun's core temperature is reckoned to be 15.6 million K, so if the value of T predicted is '2', then this means the predicted core temperature of the star is 30.2 million K, and so on.

Variable beginnings

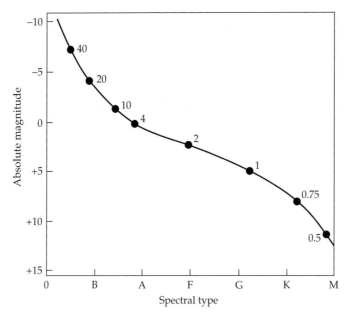

Figure 6.9 Stellar masses on the main sequence of an H–R diagram. The numbers are multiples of the mass of the Sun.

This result is intuitively not surprising. After all, the greater weight of material pressing on the core of a more massive star will mean that the core must be hotter if the gas and radiation pressure is going to be enough to support that greater weight. A hotter core will result in a much greater outpouring of energy (remember that nuclear reaction rates increase sharply with temperature) and you might expect the photosphere of the star also to be hotter in the more massive examples (remembering also Stefan's law) – and you would certainly be right when considering main sequence stars.

The vastly greater outpouring of energy with increasing temperature also means that massive stars are more 'puffed up' than less massive ones. So, stellar densities actually decrease with mass. This is true even of their core densities.

So it is that a star's mass determines its temperature (and hence spectral type) and its luminosity. In turn, it is these quantities which define a unique position for the star on the H–R diagram. It follows that any 'hydrogen burning' star should have a position on the main sequence that is determined by its mass. This intuitive reasoning is confirmed by the astrophysicists' mathematical models, which define a *mass–luminosity relation* for main sequence stars and give rise to the predictions of masses related to positions on the main sequence as illustrated in Figure 6.9.

While our Sun has an absolute magnitude of about $+5^m$ and is of spectral type G. An 8 solar mass star has an absolute magnitude of about -2^m and is of spectral type B. A one eighth solar mass star has an absolute magnitude of about $+14^m$ and is of spectral type M.

6.5 Stellar adulthood and stability

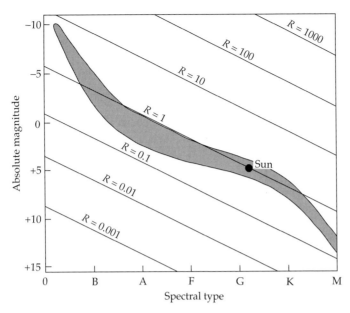

Figure 6.10 Stellar radii on an H–R diagram.

As an aside, the simplest expression of the mass–luminosity relation is of the form $L = M^\alpha$ but the value of α actually varies with mass, being around 3.8 for stars of solar mass and decreasing with both increasing and decreasing mass to values of just below 3 and 2 for the heaviest and lightest stars, respectively. As before, the values of L and M are in units relative to the solar values.

One theoretical result which covers all stars and translates directly to the H–R diagram is that stellar radii show on the diagram as a set of diagonal lines (see Figure 6.10). Notice how stellar radii increase towards the upper right, which is the area inhabited by red stars of luminosity classes III (the giants), II (the bright giants), and I (the supergiants).

6.5 Stellar adulthood and stability

It is the physical (and hence mathematical) interdependent relationships between the mass of a star, its radius, and its core and surface temperatures which gives rise to its stability. Indeed, the nuclear processes that occur in the cores of main sequence stars can even be said to be self-stabilising.

If any disturbance were to cause a star to contract slightly, this would cause the temperature to increase in its core. The nuclear reaction rate would increase in response, further boosting the temperature in the inner regions of the star. The resultant increase in radiation and hydrostatic pressure would then tend to oppose the original contraction, and even reverse it. As a result the core temperature would fall once more and ... you get the idea of course.

Most stars spend by far the largest part of their active lives on the main sequence but just how long a star remains there depends upon its mass. This is because the rate of conversion of hydrogen to helium is highly temperature dependent and, as we have seen, the core temperature is linked to a star's mass.

In the last chapter I outlined the processes that have kept our Sun, a typical main sequence star, shining for 4.6 billion years and will do so for about as long again in the future. For such a star it is the proton–proton reaction which is the dominant energy-producing reaction. This is also true of less massive stars, though the rate of consumption of their hydrogen fuel is very much less and so they last very much longer, even though there is less fuel to start with. Mathematical models predict that our Sun should last for between 9 and 10 billion years on the main sequence, while the one eighth solar mass star we considered earlier would have a main sequence lifetime of the order of 75 billion years!

The higher core temperatures of more massive stars mean that the balance of energy-producing nuclear reactions shifts until at a core temperature of roughly 20 million K the carbon cycle takes the lead rôle. This is the case for stars only a little more massive than the Sun.

Energy production rates for the carbon cycle increase even more rapidly with temperature than is the case for the proton–proton cycle. This is why it takes over the lead from the proton–proton cycle. It also results in the more massive stars gobbling up their nuclear fuel voraciously. Our previous example of an eight solar mass star will only last on the main sequence for about 30 million years, as opposed to the billions of years for less massive stars. It is a very good thing for us that our Sun is no stellar heavyweight!

Knowing the luminosity of a star and its radius allows its photospheric temperature to be determined since these values allow the energy flux per unit area of the photosphere to be calculated. Stefan's law then provides the value of temperature of the photosphere that corresponds to this value. Being pernickety, I should say that this temperature value is properly defined as the *effective temperature* – the temperature a black body would have to have in order to emit the same power per unit area of surface.

As the star converts hydrogen to helium, the helium gradually settles to the centre of the core. This causes a slight progressive restructuring of the star as time goes on. The core's density increases so it contracts a little and its temperature increases slightly. This also increases the temperature of the whole body of the star by a little, causing the star to brighten and its radius to increase. The increase in surface area of the star has a marginally greater negative effect than the increase in power output of its core, so the stellar photosphere cools slightly. On the H–R diagram this is reflected by the star moving upwards a little as well as slightly to the right. The star is at this point still inside the main sequence band. It is this very slow evolutionary change that happens to all stars which gives the main sequence band the width it has on the H–R diagram.

When most of the hydrogen in the core has been converted to helium, the core eventually develops a thinning 'hydrogen burning' shell around a hot but

6.6 The fate of a low-mass star

Figure 6.11 Planetary nebulae imaged by the author with the STV camera and 0.5 m telescope of the Breckland Astronomical Society: (a) The Ring Nebula, M57; (b) The Dumbbell Nebula, M27.

inert helium core. At this point the power output from the nuclear reactions is dropping away. This results in first the core, then the star as a whole, beginning to shrink.

I should say at this point that astrophysicists' ideas are not as certain as you might read in many accounts. What happens next, and particularly the timescales involved, are affected by a number of factors which are still rather poorly determined. Just one example is the matter of possible convection of the star's bulk of hydrogen into the core. This may well prolong the main sequence phase by a little and affect the subsequent evolution. There are other factors. With that cautionary note in mind, I can present the *likely* scenarios for stars of low, medium and heavy mass in the following sections.

6.6 The fate of a low-mass star

The term *low-mass star* denotes one which has a mass of less than half that of our Sun. Such stars enjoy very long lives on the main sequence as relatively dim objects of spectral type M. When the hydrogen burning ceases, the star radically changes its structure. The now helium-rich core rapidly shrinks, pouring gravitational potential energy into the star's outer envelope. It is thought that much of this envelope is then puffed away into space, to form a *planetary nebula*.

Planetary nebulae take the form of shells of gaseous material surrounding the central star. The most well-known example of this phenomenon is the Ring Nebula, in the constellation of Lyra, shown in Figure 6.11(a). Other examples are shown in Figure 6.11(b) and Figure 6.12(a), (b) and (c).

In these beautiful objects short-wave electromagnetic radiations from the central star cause the distended shells of gas to fluoresce. In typical examples the gas expands away from the central star with velocities from a few kilometres per second to a few tens of kilometres per second. As the gas expands so the

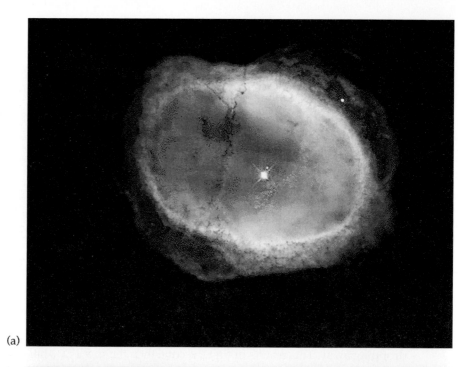

(a)

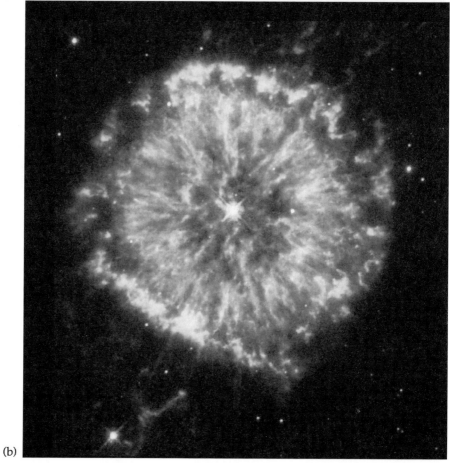

(b)

Figure 6.12

6.6 The fate of a low-mass star

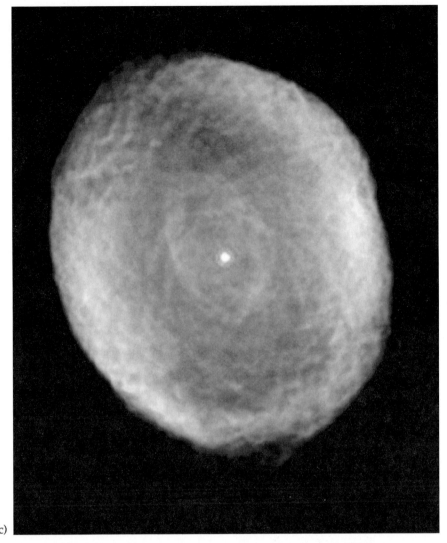

(c)

Figure 6.12 (*cont.*) Planetary nebulae photographed by the Hubble Space Telescope: (a) NGC 3132; (b) NGC 6751; (c) IC 418, often referred to as 'The Spirograph Nebula'. Courtesy of NASA and the Hubble Heritage Team (Aura/STScI).

nebula thins and eventually becomes invisible. For this reason all the planetary nebulae that can be seen have diameters not much more than a light year.

While the outer layers of the star are being ejected, the bulk of its mass shrinks under gravitation, now that the nuclear fires are extinguished. The great weight of the star crushes down on the material within it. Eventually the enormous pressures cause the boundaries of the atoms to break down, allowing the matter to reach incredible values of density.

A normal atom contains a tiny nucleus, with a diameter typically of around 1×10^{-14} m. The electrons 'orbiting' the nucleus occupy a spherical region roughly 1×10^{-10} m across. Thus an atom is mostly empty space! If someone were to throw a brick at you, it might be difficult to imagine that most of what has hit you is a vacuum but this is, in fact, true! Of course, in the high-temperature environment inside a star atoms cannot survive intact. The electrons are stripped from the nuclei to form a plasma of interspersed electrons and nuclei.

Under the enormous pressure produced by its own weight, the dead star's electrons and nuclei are forced together very much more closely than is possible even in normal atoms. The star is transformed into a superdense body known as a *white dwarf*. In terms of quantum mechanics, the electrons now occupy energy levels relative to the star, rather than relative to atomic nuclei. Moreover, in a white dwarf all the lowest energy levels are filled up with few or no spaces left. Physicists say that matter in this condition is *electron-degenerate*. The collapse of the dead star is then halted by the pressures caused by this quantum-mechanical limitation, which we call *electron degeneracy pressure*.

In its final state the white dwarf's radius is about one hundredth that of the Sun (little bigger than the Earth) and its density is around 3×10^8 kg m^{-1}, 300 000 times that of water. In fact, the more massive white dwarfs have smaller radii and so even higher densities. They are peculiar objects. Their central temperatures are only around 1 million K and this temperature varies little throughout the major part of their interiors. We cannot think of white dwarfs as being gaseous. They are more like huge crystals, gradually cooling. Only quantum physics can adequately describe matter in this state. The surfaces of white dwarfs are very hot, around 100 000 K for newly formed white dwarfs. They rotate very fast, taking perhaps only 10 seconds to turn once on their axes.

The magnetic field of the original star is concentrated to a strength millions of times greater than that of our Sun's field. The transition from a low-mass main sequence star to a white dwarf takes only a few thousand years, and so astronomers have only a sketchy idea of the path such a star takes across the H–R diagram. Still, over a hundred of the final white dwarfs have been identified.

The central stars we see in many planetary nebulae are white dwarfs. In fact, white dwarfs must be common. Astronomers estimate that about 10 per cent of the stars in our Galaxy are these stellar relics, though their low luminosities make them difficult to detect. Over the course of aeons white dwarfs gradually cool, eventually disappearing from sight as they radiate the last remnants of their thermal energy into space.

The central star of the Ring Nebula, shown in Figure 6.11(a) is very blue (and therefore very hot) and is thought to be on the verge of becoming a white dwarf. Interestingly, there is considerable disagreement between amateur astronomers about the size of telescope needed to show M57's central star. My personal opinion is that it might well be variable in its brightness (I am certainly not the first to think this), undergoing eruptions while it is starting the process of rearranging itself into a white dwarf. However, a few photometrists have investigated and found no evidence of variability.

The fact that it is immersed in nebulosity makes this a particularly difficult target for practical study. Matters are not helped by the fact that the star is of the fifteenth magnitude. You might care to keep a watch on it if you have the provisions for photometry. You will find a very convenient sequence chart given on page 102 of the September 2001 issue of *Sky & Telescope* magazine. Please, though, do be careful and rigorous in your investigation and beware of making false claims. Remember, the uncertainty of any reading is at least equal to the square root of the count, even when all other parameters are ideal. Fainter objects demand longer integration times in order to get the count up to reasonable levels.

6.7 The evolution of a star like the Sun

Considering now stars of mass between half and twice that of the Sun (main sequence absolute magnitudes between $+7^m$ and $+2^m$ and spectral types ranging from A to K), their nuclear reactions can go beyond the stage of converting hydrogen to helium. As with the less massive stars, a helium core develops with the 'burning' hydrogen confined to a shell around the core. When the nuclear reactions eventually dwindle the core contracts and becomes electron-degenerate as before. The consequent release of gravitational potential energy causes the outer regions of the star to swell up and redden. The star becomes a red giant.

When this happens to our Sun, in about 5000 million years time, it will attain a radius about 30–50 times its present size and a surface temperature of about 3500 K. Some popular books suggest that at this stage the Sun will swallow up the orbits of the inner planets, including the Earth. This is not so, although Mercury might well be engulfed. However, it is true to say that the Sun's luminosity will increase to around 500–1000 times its present value. It will probably sustain that for a few hundred million years. As a result the Earth will be stripped of its atmosphere and the oceans will boil away, leaving the Earth a scorched and dead globe.

The degenerate helium cores of these stars can contract under their own weight until they reach higher temperatures than those in their less massive counterparts. When the temperature reaches about 100 million K the fusion of helium to make carbon and oxygen is possible. However, before these reactions can begin the core has to expand to remove (in its outer parts at least) degeneracy. This only happens at a higher temperature. When that higher temperature is reached, helium 'burning' begins violently. This produces a surge in luminosity lasting a few centuries, known as the *helium flash*.

It is at this time that a star such as our Sun will give one final heave and puff out its tenuous outer layers so that, this time, all the inner planets are indeed engulfed. As this happens a star like our Sun will lose a large fraction of its mass, which will probably go to form a planetary nebula.

Variable beginnings

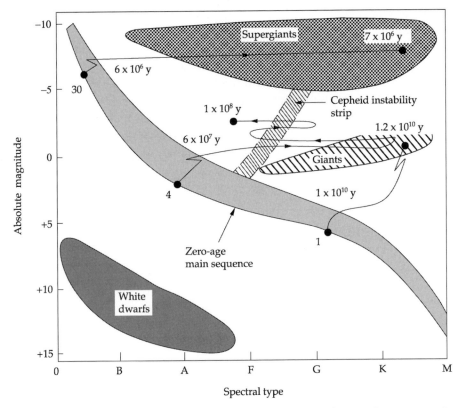

Figure 6.13 Examples of stellar evolutionary tracks on an H–R diagram for stars starting out from the ZAMS at 1, 4, and 30 solar masses.

After the helium flash (or flashes, as it can happen several times) the star will settle into a new equilibrium condition. Around the inert core will lie a zone of helium burning, and above this a zone of hydrogen burning. The outer layers will gradually expand a little further and the overall luminosity of the star will increase to over a thousand times its main sequence value. The timescales of these events are a little uncertain, as are the precise details such as the amount of mass lost by ejection.

Eventually the helium and hydrogen fusion reactions die. The core of the star then evolves to become a white dwarf, the outer envelope having been lost to form a planetary nebula, as already mentioned.

The star's changes in luminosity and effective temperature as it evolves can be plotted as a track across the H–R diagram (see Figure 6.13). As the star changes from being a red giant to a white dwarf it passes through regions of stellar instability. At these times the star turns into a variable. There is more about these episodes later in this book, beginning in the next chapter. Theorists think it likely that the final white dwarf will have only two thirds of the initial mass of the star, the rest being lost through stellar winds and planetary nebula forming outbursts.

6.8 The evolution of a massive star

Stars of mass greater than about 2.2 times that of the Sun but less than 5 solar masses (main sequence absolute magnitudes ranging about $+1^m$ to -1^m, spectral types A–B) do not develop largely degenerate cores. In these stars the helium-burning phase begins steadily and so they do not show the helium flash phenomenon. Instead, they exhibit complex looping paths across the red giant region of the H–R diagram as they undergo successive stages of core and envelope contraction and expansion. Eventually they shed a great deal of their initial mass in repeated outbursts (forming planetary nebulae) and the stellar remains become white dwarfs.

Astrophysicists are at their most uncertain when they try to predict the fates of stars of still greater mass. The amount of mass lost through stellar winds and planetary nebulae-forming outbursts is crucial in determining what happens to a star, and this is a very poorly understood factor. Certainly, massive stars use up their fuel reserves voraciously. A star of 30 solar masses will remain on the main sequence for only 6 million years. During this stage the star is a giant of spectral type O, with a luminosity of more than 100 000 times that of our Sun (absolute magnitude about -7^m) and a radius 7 times greater. At the end of its main sequence life, the core of the star contracts and the envelope expands as a helium-burning shell ignites.

The subsequent path of the star across the H–R diagram is much simpler, being virtually a straight line with no loops. The star passes through the Cepheid instability strip, for a time becoming a variable. About a million years after it left the main sequence the star has increased its diameter from its initial 7 times that of our Sun to around 1000 Sun diameters. The star has then become a red supergiant. At this stage the star has an inert core rich in carbon, nitrogen, and oxygen. The helium-burning zone envelops the core, in turn overlaid by the hydrogen-burning zone, then the outer layers of the star.

As the reaction rates in the hydrogen and helium zones dwindle, so the core shrinks and heats even further. Nuclear reactions begin in the core, synthesising progressively heavier elements, such as magnesium, sodium, aluminium, and neon. As each burning stage begins, so the core shrinks and heats up even more. Iron ultimately builds up in the core of the star, being synthesised from the fusion of silicon, and this leads to the star's downfall. The fusion of the lighter elements into heavier ones results in the release of energy but to induce iron to build up more massive nuclei energy has to be put in. Once formed, the iron remains inert. The iron core contracts until its temperature reaches 5000 million K, when photons of electromagnetic radiation have energies sufficient to begin breaking the iron down into less massive elements, such as helium. As a result the core very rapidly loses thermal energy, triggering a catastrophic collapse. This takes only a few seconds and the resulting implosion of the star's outer layers and the sudden release of energy-carrying particles, called neutrinos, cause runaway nuclear reactions in these outer layers.

Variable beginnings

The result is a tremendous explosion, a *supernova*. Most of the star's material is blasted away into space in a spectacular outburst, but the kickback on the star's iron core causes it to collapse to a superdense object known as a *neutron star*. However, here is not the place to consider supernovae and neutron stars. More about those later in this book. It is, though, worth noting here that supernovae are responsible for enriching the interstellar medium with elements of atomic mass greater than that of helium. In addition, the extreme conditions occurring during a supernova blast are responsible for synthesising all the elements with atomic masses greater than that of iron.

Stars that are subsequently born from this enriched material show the evidence in their spectra (still, though, the major factor affecting stellar spectra is temperature). Hence the younger a star is, the greater the proportion of heavy elements it will contain. Astronomers have consequently divided stars into two main categories according to their compositions.

Population I stars are the youngest and so have the greatest 'heavy' element abundances. Our Sun is such a population I star. *Population II* stars are the oldest, formed at a time before many supernovae had exploded their processed material into space. Consequently they are very heavy element depleted. I should mention that there are a small number of stars known which are extremely 'metal poor' (remember, stellar astrophysicists tend to call all elements apart from hydrogen and helium 'metals'!) and whose low masses mean that they have survived even longer than most of the population II stars. A star of mass less than eight tenths of that of the Sun can theoretically still be shining even if it was formed right back at the formation of our Galaxy, some ten billion years ago. Astronomers tend to refer to these as *population III* stars.

Next we can look more closely at why previously stable stars can become variables at some stages in their lives.

Chapter 7
Clockwork pulsators

After its unstable birth, a star will settle down to a relatively sedate existence on the main sequence for perhaps nine tenths of its total lifetime. During this period it will remain in hydrostatic equilibrium while it is steadily fusing hydrogen into helium in its core.

As described in the last chapter, it is when the hydrogen fuel becomes depleted that a star is forced to rearrange its structure and its size. The star's luminosity and spectral type change as a consequence of this. These macroscopic changes show up as evolutionary paths on a H–R diagram. In many cases these paths take the star into regions of the diagram that are the domain of the known variable stars. It then takes on the characteristics of the variable group it is visiting – but what is going on during these episodes to cause the stellar variability?

7.1 A pulsating menagerie

The Dutch teenager John Goodricke discovered the variable natures of three stars in the eighteenth century. One of these was δ Cephei, in 1784. Continued observations showed that δ Cephei varied in a wonderfully predictable way. In 1894 A. A. Belopolsky investigated this star by taking a series of spectra of it as it varied though its regular cycle of 5.37 days. He found that there were changes to the spectra arising from changes in the photospheric temperature of the star from about 5500 K to about 6500 K. The maximum temperature occurred almost at maximum brightness of the star (just a few hours earlier) but the minimum temperature occurred about a day before minimum brightness. In other words, the star's photospheric temperature started to increase about a day before its brightness began to surge upwards. The fall in brightness was preceded by a few hours by the beginning of a downturn in the photospheric temperature.

The most significant of Belopolsky's findings, though, was that the spectral lines in the star showed a periodic shift back and forth that was approximately

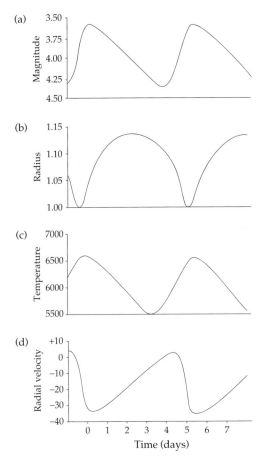

Figure 7.1 The behaviour of δ Cephei: the variations with time (a) the visual magnitude; (b) the radius of the star in units of radius/minimum radius; (c) the temperature of the star (measured in K); (d) the radial velocity of the star measured in metres per second. All the graph are matched to the same times in the cycle of this pulsating star.

in antiphase to the brightness variations. In other words the star appeared to be first approaching and then receding from us as the cycle progressed! To refine this statement slightly I should mention that the star has a mean recessional velocity of about $20\,\mathrm{m\,s^{-1}}$. So, the observed velocity actually varies from an approaching $4\,\mathrm{km\,s^{-1}}$ to a recessional $36\,\mathrm{km\,s^{-1}}$. Figure 7.1 should make all this clear.

A real bulk motion of the star arising from the star orbiting any hypothetical invisible companion was ruled out on the grounds that it did not explain the temperatures variations. Knowing that the star's spectrum is generated by its photosphere and chromosphere, it was realised that it is just this part of the star which is cyclically approaching and receding.

The inevitable conclusion is that the star is cyclically expanding and contracting, almost as if it were breathing. In fact, δ Cephei changes its radius by about 14 per cent as it first swells and then contracts. The star achieves maximum size roughly half way through its minimum-to-maximum brightness cycle (see Figure 7.1).

A great many stars vary their brightnesses because of such bulk pulsations. Indeed, this is the reason why *most* variable stars are variables. Roughly two thirds of the stars catalogued in the GCVS are pulsating variables. Mind you, they are very far from being alike. They vary greatly in all their macroscopic characteristics (excepting that they tend to occur in well-defined groups on the H–R diagram). They also vary in their amplitudes and periods of oscillation and in the precise way the brightness varies with time (as revealed by the shape of their light-curves). Some are extremely regular, repeating their cycles identically time after time. Others are much less regular.

Consequently if you consult the GCVS file on the accompanying CD-ROM you will find 35 different (including the subtypes) categories of pulsating variable star listed. These are: *ACYG, BCEP, BCEPS, BLBOO, CEP, CEP(B), CW, CWA, CWB, DCEP, DCEPS, DSCT, DSCTC, L, LB, LBV, LC, M, PVTEL, RR, RR(B), RRAB, RRC, RV, RVA, RVB, SRA, SRB, SRC, SRD, SXPHE, ZZ, ZZA, ZZB*, and *ZZO*.

As an example, the stars that behave like δ Cephei are the ones classified DCEP by the GCVS (DCEP stands for 'Delta Cephei'). Inevitably, and perhaps confusingly, there are a few other categories for pulsating variables recognised by other authorities. For instance you may come across '53 Per stars', 'UU Herculis stars', and one or two others in various references.

Let me restate a point I made earlier in this book. Be aware of the different classifications given by different authorities but do your best to stick to the GCVS Research Group's scheme. This is especially important if you are new to this subject and are still finding your way around.

Remember, you will find a concise description of each of the GCVS types and subtypes, including their macroscopic characteristics and details of their variability, on the accompanying CD-ROM. In this chapter and continuing into the next one, I shall limit myself to highlighting just a few of the main types of pulsating variable star.

7.2 The physics of stellar pulsation

Many things oscillate: flags flutter in the breeze; pendulums swing; bells ring after being struck; the reed in a woodwind instrument vibrates as air is blown across it. We have already seen how the observational evidence led to the discovery of stellar pulsation. So we know for certain that stars can also oscillate, in spite of their enormous sizes.

A characteristic of relatively simple structures, such as a pendulum or a string on a violin, is that they can only be made to vibrate at one frequency, or perhaps

Clockwork pulsators

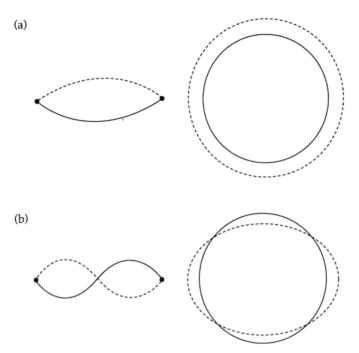

Figure 7.2 Illustrating the equivalent modes of oscillation for a violin string and a star. The solid and dotted lines represent the extreme positions of the string and surface of the star, in each case: (a) fundamental mode; (b) first harmonic.

at a number of well-defined frequencies that are relatively simple ratios of one particular 'preferred' frequency.

When the oscillating system, be it violin string or star, is vibrating with its preferred frequency it is doing so in the simplest way possible. For the string this is where just one loop is fitted in between the ends of the string. The ends of the string are positions of 'no-motion' or *nodes*. At the midway point the string is moving back and forth with its greatest amplitude. This position is known as the *antinode*. At any one instant all parts of the string are simultaneously moving in the same direction. For the star the equivalent is where the whole surface of the star is moving outwards or is moving inwards at the same instant (see Figure 7.2(a)). In either case this is the *fundamental mode* of oscillation.

As an aside, do please note that the string undergoes a *transverse* oscillation. The particles in the string move at right angles to the direction in which the wave itself moves. The particles in the star move in the same direction that the wave propagates. This is known as a *longitudinal* oscillation. Sound waves moving through air are an everyday example of a longitudinal wave motion.

The next more complicated way the string can oscillate is to have two loops formed between the ends of the violin string (two antinodes). This time while one half of the string is moving in one direction, the other half of it is moving in the other. The equivalent in the star is to have one region girdling the surface

7.2 The physics of stellar pulsation

of the star (for instance the equator) moving outwards at the same time as the other areas (for instance the poles) are moving inwards. The string and the star are now oscillating with a frequency which is higher than the first one and is known as the *first harmonic*. This mode of oscillation can also be referred to as the first harmonic. The situation is illustrated in Figure 7.2(b).

Increase the number of loops on the string and you have the second harmonic, third harmonic – and so on – modes of oscillation. Increase the number of separate patches of the stellar photosphere which are simultaneously moving outwards while adjacent patches are simultaneously moving inwards and you have the (admittedly more complicated looking) equivalents for the star.

As we shall see, many variable stars pulsate in modes higher than their fundamental. Some even pulsate in ways which are mixtures of the different modes! We then speak of the oscillations being a mixture of the fundamental mode and *overtones*.

The oscillations of a simple pendulum are easy to understand. A pendulum bob on a cord that is hanging straight down and motionless will remain in that state until something pulls it aside and then releases it. When the bob is displaced sideways it is a component of the gravitational force which acts in such a direction in order to try to restore the pendulum to its previous position. The more the bob is pulled sideways, the greater is this restoring force. When released the bob accelerates under the influence of this force. This sideways component of force lessens and becomes zero as the bob regains its original position.

However, there is a complication. Momentum carries the bob beyond this point. Consequently it now swings out in the other direction. Once again the restoring force grows and the bob first slows, momentarily comes to rest, and then begins to accelerate back towards its rest position. Again momentum makes it overshoot. Off we go again . . .

In the simplest terms what happens in a pulsating star is not too different. If something causes the star to expand beyond its normal size – the size set by the condition for hydrostatic equilibrium (we will worry about the cause shortly) – gravity will act in order to pull the star back in. However, like the pendulum bob, the moving material of the star will have momentum. It will overshoot and the star will contract below its normal size. No longer being in hydrostatic equilibrium, the building gas pressure will first slow, then stop, the contraction and subsequently force the star to expand once more. It will overshoot and the cycle will repeat.

With the simple pendulum it was just a component of gravity acting on the bob, first in one direction and then the other. In the case of the star it is the cyclic predominance of gas pressure wanting to make the star expand followed by gravity wanting to make it contract. In either case the result is oscillation.

Left to themselves, oscillating systems gradually lose energy and eventually cease oscillation. Usually this is due to frictional forces – for instance in the case of the simple pendulum there is some stiffness in the cord as well as a damping force caused by the motion of the bob and cord through the air. A whole variety of differing damping forces (the Joule–Kelvin effect causing the gas to lose energy to radiant heat as it is compressed, turbulence within the star, magnetic field interaction with plasma, etc.) would in a short time stop a star oscillating.

So, there has to be some factor operating to supply energy to the oscillations in order to keep them going. Sir Arthur Eddington was the first to make real headway in explaining the mechanisms driving stellar pulsation. By 1917 he had derived a predictive mathematical relationship that mechanically explained a star's period of pulsation.

He also was able to show that the amplitude of oscillation was greatest at the surface of the star but very rapidly decreased towards the centre of the star. This is because of the steep increase in density towards the centre of the star. Eddington deduced that the cause of the oscillation probably lay nearer the surface of the star than the core for this reason. To take a somewhat dubious analogy, think where it is most efficient to apply a periodic push to a child's swing in order to keep the amplitude of the swing the same – and why.

Eddington proposed that there was some form of 'heat valve' situated somewhere in the outermost region of the body of the star. When the valve was closed heat energy would build up below it, so causing a rise in temperature in the immediate 'subvalve' layer and a subsequent surge in gas pressure in this layer. The 'subvalve' layer would expand as a result, pushing upwards all the material above it. The momentum of the upward moving material would cause the expansion to overshoot. The valve would open, heat energy would escape, and the previously expanded layer would cool and contract. The uplifted material would fall back. The contraction would overshoot, the compressing gas would heat up, exacerbated by the now reclosed valve. The cycle repeats again – and again – and again . . .

This was a brilliant piece of thinking on Eddington's part. He wondered whether under certain conditions thin zones of stellar material can become opaque to the passage of radiation, so explaining these hypothetical heat valves. It turns out he was correct about this, too, although not enough was known at the time for him to identify the causative agent definitely.

We now know that under certain conditions stellar gases can become *partially ionised* (some but not all atoms are stripped of their electrons). In this state the gas will contain a very high proportion of un-ionised but highly excited atoms – ones with their electrons in high quantum energy levels. These atoms readily absorb the extra energy needed to totally free their electrons. Hence the gas as a whole is hungry for radiant energy and heavily absorbs it.

For an effective heat valve to occur in a star the temperature and pressure have to be just right. Totally ionised gas (plasma) is relatively transparent. So is totally un-ionised gas. If the pressure is low, then so is the density of material. If this is too low, then radiation can pass with only a little of it being absorbed, even if the temperature happens to be just right for partial ionisation.

7.2 The physics of stellar pulsation

When the conditions are met at all, the result is a relatively thin layer of partially ionised gas. We call this the *ionisation zone*. A characteristic of this material is that it is very opaque to the passage of radiation. You might recall from Chapter 5 that the outermost zone of the interior of our Sun has such an ionised layer, which is why turbulence takes over from radiative transfer as the main mechanism transporting energy from the interior to the photosphere.

Another characteristic of ionised material is that if the gas in the zone is compressed its opacity increases even further. This is exactly the characteristic needed to satisfy the requirements of a stellar 'heat valve'. Even better, while compressing, the ionisation zone is voraciously taking in energy as more material is being ionised. This helps the contraction along by removing heat energy from the adjacent layers of the star. When the ionisation zone is expanding it is de-ionising and then losing energy to the adjacent layers of the star. This gives a further push to the expansion.

The partially ionised gases in our Sun are immediately below its surface and so there is no significant material above it to force upwards; nothing for the heat valve to operate on. The ionisation zone has to lay deeper within the star for the star to pulsate. The higher temperatures of the still deeper layers mean that here the gas is totally ionised and so is relatively transparent to radiation.

To take one instance, stars that are just beginning to evolve away from the main sequence with photospheric temperatures around 7000 K (a little hotter than our Sun) develop a zone a little below the photosphere where the conditions are just right for the mix of hydrogen and helium to be partially ionised (it is the helium which is the 'active ingredient' in this case). The layer is at a sufficient depth to control a large mass of material above it but not so deep that the density makes the amplitude of oscillation too small (remember the most efficient place to push a child's swing). Consequently, these stars pulsate.

If you take a look at a H–R diagram (such as that shown in Figure 6.1) you will find that this corresponds to stars of spectral type F, right where the Cepheid instability strip comes close to the main sequence. This is an effective verification of Eddington's ideas, which are still accepted as explaining the basic mechanism of stellar pulsation.

Certain locations on the H–R diagram correspond to stellar macroscopic characteristics (and thus internal structures) which lend themselves to partially ionised zones in the stellar interiors at just the right depths to drive a star into instability. The star then oscillates at either its fundamental or higher harmonic frequencies, or a combination of these.

The star's pulsations are maintained by the ionisation layer, acting as its own natural heat valve, until evolution changes the star's macroscopic characteristics enough to render this layer ineffective (by it becoming too near the surface or too deep, or if it is simply disrupted by convection or destroyed by an unfavourable temperature and/or temperature gradient).

Let us now examine a few specific types of pulsating variable star.

7.3 CEP (Cepheid) and CEP(B) stars; DCEP (Classical Cepheid) and DCEPS stars; CW (W Virginis), CWA and CWB stars

Cepheid variable stars (generally denoted *CEP*) are a class of fairly massive stars, typically 3–9 times the mass of our Sun. They are very brilliant stars of luminosity classes I and II (about 500 to 30 000 times the luminosity of our Sun) of spectral type F. Their pulsation periods range from about 1 day to about 50–60 days in the main, though a small number can have periods up to about 200 days. Their brightness amplitudes can range from a couple of magnitudes down to a couple of hundredths of a magnitude.

CEP stars are single-mode oscillators but *CEP(B)* stars pulsate in a mixture of two modes blended together. Cepheid variables that are similar to δ Cephei are denoted as *DCEP* stars. These are reckoned to have only recently left the main sequence and have moved into the Cepheid instability strip for the first time.

Those having brightness amplitudes of less than $0^m.5$ are denoted *DCEPS*. Our familiar, friendly, Pole Star (Polaris aka α Ursae Minoris) is of this type. Its apparent magnitude varies from $1^m.95$ to $2^m.05$ over a period of 3.97 days.

In 1912, five years prior to Eddington's research, Henrietta Leavitt had identified a number of Cepheid variable stars in the Small Magellanic Cloud (SMC), which is one of the satellite galaxies of our Milky Way. She found that the brightest of these stars had the longest periods of oscillation. Knowing that all these Cepheids are at approximately the same distance from us (all being inside the SMC) she could relate the mean apparent magnitudes of the stars to their mean absolute magnitudes, and hence to their luminosities. After continued study Henrietta Leavitt was able to determine that there was a *period–luminosity relation* (sometimes known as a *period–luminosity law*) for the Cepheid variable stars.

I should mention that there is also a slight dependence on a star's photospheric temperature, since this is also related to luminosity. Hence stars with the same luminosities and slightly different colours (as far as allowed by the width of the region on the H–R diagram) will have slightly different periods. This produces some scatter in the period–luminosity relation.

In his own studies, Eddington did not stop at qualitatively explaining why a star might pulsate. He did his best also to explain the observed period–luminosity relation. In the case of a pulsating star the enormous densities at its centre effectively make this position a node. The photosphere of the star is free to oscillate with the maximum possible amplitude, and so corresponds to an antinode. Remember, in the case of the star oscillating in its fundamental mode all the mass motions are longitudinal and so are in the radial direction. So Eddington reasoned that the natural period of oscillation for a star would be equal to the time taken by any hypothetical vibrational disturbance created at the photosphere to travel down to the core, then to be reflected back outwards from the core and finally to reach the photosphere once again.

7.3 Cepheid variable stars

Of course, in a star of large radius the disturbance can be expected to take longer to go from the photosphere to the core and back again. So, bigger stars should have longer periods of oscillation. However, the speed of the disturbance is not constant as it travels through the stellar interior. Its speed actually increases with density. Hence the disturbance is moving fastest in the deep regions near the core of the star, even though the amplitude of motion is here close to its smallest value.

Actually, big stars *do* oscillate with longer periods than small stars but Eddington had to take everything he could into account in developing a predictive mathematical relationship. He found that for a star oscillating in a given mode the period is inversely proportional to the square root of the average density of the star:

$$\text{period} \propto \frac{1}{\sqrt{\rho}}$$

As well as providing data for Eddington's studies, Henrietta Leavitt's work bore yet more fruit. Ejnar Hertzsprung built on her discovery by calibrating the brightness scale using the nearest Cepheids whose distances were known from parallax measures. As a consequence, observations of a Cepheid variable could be used to determine its pulsation period. This, by the period–luminosity law, determined its absolute magnitude. *Having established its absolute magnitude and measuring its apparent magnitude, the star's actual distance easily follows* (see Section 1.2).

Used in this way Cepheid variable stars are *standard candles* – objects whose true brightnesses are known (or can be easily determined) and so whose distances can be found by making a comparison with their apparent brightnesses. DCEP stars are not the only standard candles used by astronomers but they were the first and they provided the first link in a chain of distance determining methods between parallax measures and those other methods that work for distances far beyond the range afforded by parallax.

In 1923 the most powerful telescope on Earth was the famous '100 inch' (2.54 m) reflector at Mount Wilson. Edwin Hubble used the telescope in that year to identify Cepheid variable stars in the spiral nebula known as M31. In doing so he was able to establish a preliminary distance for this nebula of three quarters of a million light years. Harlow Shapley had, by then, established the size of our own galaxy and so Hubble's discovery proved that M31 (and subsequently all similar objects) lay well beyond the confines of our own galaxy. Suddenly the Universe became a very big place!

The reason I said that Hubble's distance to M31 was 'preliminary' was that it turned out that the Cepheids that Hubble picked out were not the same as those we now know as DCEP stars. DCEP stars are stars of population I – and are sometimes called *population I Cepheids*, or *classical Cepheids*. The ones identified by Hubble were in fact Cepheids of population II. While the classical Cepheids

Figure 7.3 The globular star cluster, M13, imaged by the author using the STV camera and 0.5 m reflector of the Breckland Astronomical Society.

are fairly massive stars, the population II Cepheids are actually much more evolved stars with lower masses. They are often known as *population II Cepheids*, or *W Virginis* stars (*CW* is the modern abbreviation for this type). He found them in the halo of M31, whereas DCEP stars are most likely to be found in the disk of the host galaxy. CW stars are most common in *globular star clusters* (see Figure 7.3). Globular star clusters are themselves a component of the galactic halo. DCEP stars follow a different period–luminosity law to the CW stars and are typically a couple of magnitudes brighter than a CW star of the same period.

Hubble consequently underestimated the distance of M31. It was then refined to about one and a half million light years, and later to 2.2 million light years. That figure was accepted for most of the latter half of the twentieth century but it has since been revised again, owing to a critical recalibration of stellar distances thanks to the Hipparcos mission. We now know M31, the Great Galaxy in Andromeda, to be 2.9 million light years (890 kiloparsecs) away.

It was Walter Baade who discovered the distinction between DCEP and CW stars in the late 1940s. CW stars are often subdivided into *CWA* and *CWB* stars, those of periods longer than 8 days, and shorter than 8 days, respectively.

It is hardly possible to overstate the importance of these 'standard candle' variable stars and the work of those famous pioneers who studied them and those who built on the results. We really do owe our understanding of the scale of the Universe, and all that flows from it, to them. Perversely, thanks to their well-defined light-curves and wonderful predictability, there is not a great deal of point in *us* observing them.

Still, you can find δ Cephei identified on even simple star maps and, of course, you can observe it easily with the naked eye. There is certainly no harm in using it to practice on. You will need a good run of clear nights if you are to get anything resembling a recognisable light-curve. At least this is a good lesson in why one observer's results become so much more meaningful when combined with the results of others.

Another example of a star that behaves like δ Cephei is η Aquilae. This star was first discovered to be a variable in 1874 by Edward Pigott, who was a close friend, near-neighbour, and observing colleague of John Goodricke. This Cepheid variable has a range of $3^m.5$–$4^m.4$ over a period of 7.2 days, so it is also easy to find on a star atlas and is easy to follow throughout its cycle by means of the unaided eye alone.

7.4 RR (RR Lyrae), RR(B), RRAB, and RRC stars

I mentioned that CW stars are most often to be found in globular star clusters. They are old stars, as are the vast majority of stars in these clusters. Much more commonplace in these giant stellar swarms are another class of pulsating variable: the *RR Lyrae* stars. These 11–13 billion year old geriatrics are a little less massive than our Sun but are puffed up to about 5 times its radius. They have spectral types in the range of A–F and so are a little hotter than our Sun. Consequently they are rather more luminous, usually between 40 and 90 times the luminosity of our Sun.

Statistically they have a well-defined absolute magnitude averaging about $+0^m.5$. So, if astronomers identify and measure the brightnesses of a collection of these stars in each particular globular star cluster, the distance modulus of the cluster (and hence is distance) can readily be found. Harlow Shapley capitalised on this fact in the 1920s in order to derive the distances of the globular star clusters that form a framework around our own Galaxy. In the process he was able to coarsely define the overall shape and size of our Galaxy and even define where the centre of it lay as well as our own location within it.

RR Lyrae stars derive their energy from fusing helium in their small inner cores (via the *triple-alpha process* whereby three helium nuclei are fused to make carbon) and from the fusing of hydrogen (via the CNO cycle) in a shell surrounding the inner core. The bulk of the mass of the star lies within this region, which defines the core of the star. In this state the core is producing much more power than it did in the main sequence phase which is why the outer body of the star is both swollen and hot.

However, the star cannot maintain this for more than a few hundred million years. This is very much the star's final sprint to the finish. It will end with the star's core collapsing to electron degeneracy, envelope expansion, one or more helium flash episodes, and the formation of a planetary nebula. The star will end as a white dwarf gradually cooling and fading into obscurity at the end of its long life.

RR Lyrae variables pulsate with periods commonly ranging from about a quarter of a day to slightly over one day and amplitudes ranging from $0^m.2$ to about 2^m. There are a few rare examples known that have periods as long as 2 days but there is a strong statistical clustering around a period of 0.5 days. It is their very short periods that led to them being noticed by S. I. Bailey in 1895.

All the RR Lyrae variables discovered by Bailey inhabit globular star clusters, as do very roughly half of those we know today.

Those that undergo two blended pulsation modes are denoted *RR(B)*. Some RR Lyrae stars have markedly asymmetric light-curves. These are denoted *RRAB*. The GCVS Research Group recognise a further subtype: *RRC*, which they define as 'RR Lyrae variables with nearly asymmetric, sometimes sinusoidal, light curves, of periods from $0^d.2$ to $0^d.5$, and amplitudes not greater than $0^m.8$ in V'.

Some RR Lyrae stars undergo periodic changes in their light-curves. This *Blazhko effect* results from a beating between the two pulsation modes. Imagine the effect of playing two adjacent C notes on an organ when one of them is very slightly out of tune. The resulting slow modulation to the amplitude of the note you hear is just the same sort of effect.

There is little point in us observing these clockwork stars. Another put-off is the fact that as a class they also tend to be faint. I have allocated some space to describing them here because they are of great importance as standard candles and they are an important part of the story of pulsating stars.

If you do want to see, or maybe even monitor, a RR Lyrae star then the best one to choose is undoubtedly RR Lyrae itself. This star is of type RRAB and its apparent magnitude varies between extremes of $8^m.12$ and $7^m.06$, with some modulation of its range due to the Blazhko effect. During its 0.57 day cycle its spectral type changes from F7 to A5. You can find this star at $19^h\ 25^m\ 28^s$ right ascension and $+42°\ 47'\ 04''$ declination.

In the next chapter we encounter variable stars that are rather less regular than those considered in this one. In doing so we open the box to a great many stars ripe for our scrutiny.

Chapter 8
Less regular single-star variables

Not all pulsating stars are as regular as the Cepheid and RR Lyrae types. As observational astronomers looking for interesting fare we can be glad of this. I am pleased to say that I can more than make up for the dearth of worthwhile stars to observe so far with the stars described in this chapter. We start with the almost regular Mira stars and progress to stars displaying rather wilder behaviour.

8.1 M (Mira) stars

There are over 5200 definite examples of *Mira-type* variables, or *M* stars, listed in the GCVS database. This is a greater number than for any other specific type of variable. There are about another 1000 stars listed which are suspected of being Mira-type variables. They are also very well represented by their charts and light-curves on the CD-ROM that accompanies this book.

The prototype star is o Ceti. It was discovered to be variable after 1596 when Dutchman David Fabricus recorded it as a star of the third magnitude in August of that year. However, it had vanished from sight by October. Johann Bayer recorded it in 1603 when he was constructing a star catalogue. It was he who gave it the letter omicron. It vanished again soon after. In the face of the dogma about the perfect and unchanging nature of the stars at the time, it took until the late 1630s (thanks to Phocylides Holwarda – yet another Dutch pioneer in this field) for it to be accepted that this star periodically brightens and fades with a period of about 332 days. This was a momentous discovery. It was also the first *long-period* variable star to be identified as such. In 1662 Johann Hevelius published his *Historiola Mirae* in which he named the star 'Mira', meaning 'The Wonderful'.

Figure 8.1(a) shows the light-curve of the star as recorded by BAAVSS members from 1891 to 2000 and Figure 8.1(b) takes a closer look with the light-curve

Less regular single-star variables

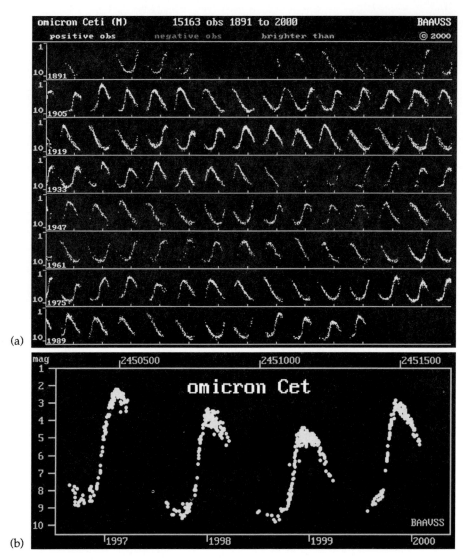

Figure 8.1 (a) The light-curve of o Ceti from 1891–2000. (b) The light-curve of o Ceti from 1996–2000. Courtesy BAAVSS.

recorded in the last four years of that period. You will notice that this star is far from behaving like the precise clockwork variables of the Cepheid and RR Lyrae types. The period is a little variable; the mean value being 332 days.

Particularly noticeable from the light-curves is the fact that the magnitudes at maximum and minimum brightness vary from cycle to cycle. In some cycles this star fails to achieve even fourth magnitude at its brightest, while on other rare occasions it can climb nearly to magnitude 1. This results in the star being visible to the naked eye for very variable durations around the times of maximum light. This can be as short as a few weeks at a time or as long as five months. For many

8.1 M (Mira) stars

months at a go the constellation of Cetus appears bereft of its 'Wonderful' star. Even small binoculars might not be enough to see the star at the times of deepest minimum light.

All M stars are highly evolved red giants of 'late' spectral types which show some spectral features in emission. They tend to have masses from a little less than that of our Sun to about $1\frac{1}{2}$ times that of our Sun. The majority are of spectral type M_e, though some are C_e (carbon-rich) and S_e (rich in zirconium oxide). The 'e' subscript denotes the presence of emission features. They are swollen red stars with photospheric temperatures of about 3700 K or less. Their diameters are reckoned to be about 200–400 times that of our Sun and vary by something like 20 per cent during their pulsation cycles.

Interestingly, as they pulsate the swollen and tenuous photospheres of these stars move at speeds faster than the natural speed of sound in the rarefied photospheric material. This must cause supersonic shock fronts to form and ripple across the stars. It is the localised heating associated with these fronts which is thought to be responsible for the emission features which show up in Mira-type spectra.

The physical conditions and the weak gravity at their photospheres cause these aged stars to be very windy – burping away significant amounts of matter into space. Indeed, they are on the verge of the rearrangements that will result in them creating their own planetary nebulae. Our Sun will become such a Mira-type star in about 5 billion years time.

Their amplitudes can be as great as 11^m and their periods range from about 80 days to about 1000 days (somewhat more or a little less in a few cases – though the GCVS Research Group imposes these limits on the M-type category). Like Mira itself, M stars have brightness amplitudes that can vary by several magnitudes from cycle to cycle and periods that can alter by as much as 15 per cent.

Convection within the star and the chaotic conditions imposed by the supersonic shock fronts are presumably what causes the variations in amplitude and period of Mira-type stars. One can imagine how the ionisation zone's position and density must be affected as a result of these factors. The result is that Mira-type stars make wonderful subjects for observation. Observing them is both interesting and valuable. A large number of them are fairly bright. That factor and their large amplitudes make them relatively easy to observe visually. Their redness can cause problems, of course, but these stars are very much the 'stock in trade' for many observers.

You will find light-curves for the following M stars on the accompanying CD-ROM. In order to be of help to you, whether or not you make use of the 'find' facility on your web browser, I give them here in the order and style in which you will find them listed in the 'lightcurves' file:

V493 Aur; V494 Aur; V854 Cas; V362 Cep; V1426 Cyg; V409 Per; V513 Per; R And; W And; RW And; R Aql; R Ari; R Aur; X Aur; R Boo; S Boo; R Cam; V Cam; X Cam; R Cas; S Cas; T Cas; W Cas; S Cep; T Cep; omicron Cet; R Com;

S CrB; V CrB; W CrB; R Cyg; S Cyg; V Cyg; chi Cyg; R Dra; R Gem; S Her; RU Her; SS Her; R Hya; SU Lac; X Oph; U Ori; R Ser; V Tau; R UMa; S UMa; T UMa; S UMi; R Vul; R And; W And; RW And; R Aql; R Aqr; UV Aur; R Boo; R CVn; U CVn; R Cam; V Cam; R Cas; S Cas; U Cas; V Cas; W Cas; SS Cas; S Cep; T Cep; omicron Cet; SU Cnc; R Com; R CrB(1); R CrB(2); S CrB; W CrB; X CrB; R Cyg; S Cyg; U Cyg; V Cyg; RT Cyg; TU Cyg; DD Cyg; V1426 Cyg; V1990 Cyg; chi Cyg; R Dra; T Dra; S Her; T Her; U Her; W Her; RS Her; RU Her; SS Her; SY Her; TV Her; R Hyd; R LMi; U LMi; SU Lac; R Leo; RS Leo; W Lyn; X Lyn; U Ori; U Per; Y Per; R Ser; R Tri; R UMa; T UMa; RS UMa; S UMi; and R Vir.

You will notice that a few of the entries are duplicated. This is where different versions of the light-curves exist (for instance a detailed short-term plot supplementing the longer-term plot). Rather than gathering these together, I thought it of most use to you if I slavishly stick to listing all entries in the order in which you will find them as you work through the document on the CD-ROM. You will come to them in blocks under specific headings so there ought to be no confusion as to the nature of the light-curve presented in each case.

I will stick to this arrangement in the rest of this book for giving the listing for all the other variable star types featured on the CD-ROM.

Of the foregoing list, the following stars have TA charts on the CD-ROM:

V493 Aur; V494 Aur; U CVn; V727 Cas; V854 Cas; V362 Cep; V409 Per; V513 Per; and V1990 Cyg.

In the 'TA chart catalogue file' you will find TA charts for the following additional Mira-type stars:

RV Cas; V666 Cas; V667 Cas; R Tau; CC Cam; V686 Mon; EO Dra; EZ Dra; and V1760 Cyg.

Referring again to the list of light-curves, the following Mira-type stars also have BAAVSS charts on the CD-ROM (given here in the style used in the file so that your browser tool can find the entries):

omicron Ceti; S Coronea Borealis; X Ophuichi; UV Aurigae; and U Leo Minoris.

In addition, the 'BAA VSS charts' file also contains the chart for R Aquarii, another Mira-type star.

As well as following the known Mira stars, you could make a significant contribution by observing stars whose attributions are presently questionable. The following stars represented on the CD-ROM may or may not be Mira variables. This has still to be decided in each case at the time I write these words:

- TAV1831+19: you will find both a light-curve and a TA chart for this on the CD-ROM.
- NSV8001 Oph: the light-curve and TA chart are on the CD-ROM.
- Z UMi: you will find its light-curve on the CD-ROM.

8.2 Semi-regular variable stars

- V418 Cas: there is a TA chart for this object on the CD-ROM.
- V409 Per: there is a TA chart for this object on the CD-ROM.

I will use this arrangement throughout the rest of this book for giving the listings for all the other variable star types featured on the CD-ROM. Where light-curves are given but no charts, at least you have the given star identified under its variable type, hopefully allowing you to obtain a chart from elsewhere (perhaps from your group's co-ordinator) if desired.

8.2 SR (semi-regular variable); SRA; SRB; SRC; SRD; and SRS stars

Look up the GCVS classification on the accompanying CD-ROM and you will find the entry for *SR* stars (meaning *semi-regular* variable stars) beginning: 'Giants or supergiants of intermediate and late spectral types showing noticeable periodicity in their light changes, accompanied or sometimes interrupted by various irregularities. Periods lie in the range from 20 days to more than 2000 days, while the shapes of the light-curves are rather different and variable, and the amplitudes may be from several hundredths to several magnitudes (usually 1^m to 2^m in V).' If that strikes you as rather similar to the specification of the Mira-type stars, then you are correct.

The main differences are that the classification of stars includes supergiants as well as giants and that the amplitudes are smaller than that ascribed to the Mira classification. While the light-curves of Mira stars are a little irregular, those of SR stars are generally even less so.

There are five subtypes. The entry from the GCVS classification scheme I have quoted from goes into much more detail but broadly *SRA* stars are giant stars that are little more regular than the giant stars of type *SRB*. *SRC* stars are supergiants, and *SRD* stars encompass giants and supergiants with an earlier spectral range (types F, G, or K) than the others (which are M, C, and S).

With radii several hundred times that of the Sun, the supergiants have even more tenuous outer layers and weaker surface gravities than do the giants, even though they are perhaps 10–30 times as massive as our Sun. Consequently the rate of mass loss can be even more appreciable, perhaps around one hundred thousandth of a solar mass each year.

Betelgeuse is an SRC star which could hardly be easier to locate with the naked eye (see Figure 1.1). It was first noticed to be variable by John Herschel in 1836. It is normally found between magnitudes $0^m.0$ and $+1^m.3$. Its brilliance does, though, mean that it is hard to find suitable comparison stars. Veteran variable star observer David Levy suggests choosing Procyon (α Canis Minoris) and Pollux (β Geminorum) as suitable comparison stars visible in the same quadrant of sky and I concur. Procyon and Pollux are of visual magnitudes $0^m.38$ and $1^m.14$. It is inevitable that there will be a difference in the altitudes of the three stars but in the absence of any closer comparison stars, you will have to do the best you can. Try to observe at a time when all of them are high in the sky to minimise the effects of atmospheric extinction.

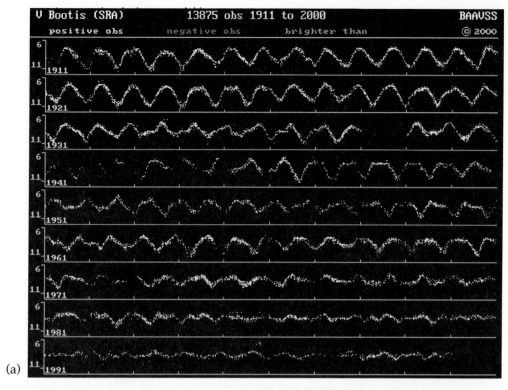

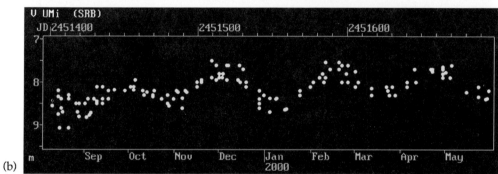

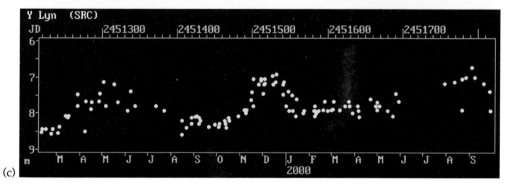

Figure 8.2 (a) The light-curve of V Boo (type SRA). (b) The light-curve of V UMi (type SRB). (c) The light-curve of Y Lyn (type SRC). (d) The light-curve of RU Cep (type SRD). Courtesy BAAVSS.

8.2 Semi-regular variable stars

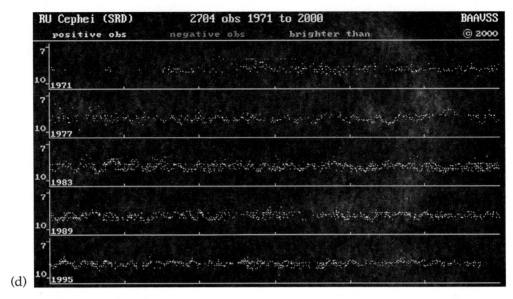

Figure 8.2 (cont.)

Betelgeuse's variations are rather erratic but have a period of about 6.4 years. This is much longer than most other SR stars – perhaps it is best categorised as an intermediate between SRC and L stars. L stars are described in Section 8.4.

SRS stars (otherwise known as *UU Herculis* stars, after the group prototype) seem to suddenly switch between modes of oscillation every so often, or even to undergo periods when the large amplitude oscillations entirely switch off!

SR stars are really fascinating to follow. Four example light-curves are reproduced in Figure 8.2. There are light curves for a very large number of SR stars on the accompanying CD-ROM. They are listed here in the order in which you will find them in the 'lightcurves' document.

V727 Cas; V386 Cep; Y Lyn (SRC); V UMi (SRB); V450 Aql (SRB); V1293 Aql; UU Aur (SRB); U Boo (SRB); V Boo (SRA); RV Boo (SRB); RW Boo (SRB); V CVn (SRA); Y CVn (SRB); TU CVn (SRB); RY Cam (SRB); ST Cam (SRB); WZ Cas (SRB); V465 Cas (SRB); W Cep (SRC); RU Cep (SRD); SS Cep (SRB); mu Cep (SRC); RS Cnc (SRC); W Cyg (SRB); RV Cyg (SRB); AF Cyg (SRB); BC Cyg (SRC); CH Cyg (one component of the binary system is an SR star); V460 Cyg (SRB); U Del (SRB); EU Del (SRB); MV Del; TX Dra (SRB); AH Dra (SRB); eta Gem (one component of this close binary system is an SRA star); X Her (SRB); SX Her (SRD); UW Her (SRB); IQ Her (SRB); OP Her (SRB); RX Lep (SRB); Y Lyn (SRC); SV Lyn (SRB); R Lyr (SRB); BQ Ori; S Per (SRC); RS Per (SRC); SU Per (SRC); AD Per (SRC); BU Per (SRC); Z Psc (SRB); TV Psc; S Sct (SRB); Y Tau (SRB); W Tri (SRC); V UMa (SRB); RY UMa (SRB); TV UMa (SRB); VW UMa; V UMi (SRB); RS And (SRA); TZ And (SRB); S Aql (SRA); V Aql (SRB);

U Boo (SRB); V Boo (SRA); RW Boo (SRB); RX Boo (SRB); S Cam (SRA); UV Cam (SRB); RW Cep (SRD); AR Cep (SRB); TR Cnc (SRB); RR CrB (SRB); SW CrB (SRB); RS Cyg (SRA); AF Cyg (SRB); St Her (SRB); IQ Her (SRB); OP Her (SRB); g Her (SRB); W LMi (SRD); SX Lac (SRD); RY Leo (SRB); Y Lyn (SRC); tau4 Ser (SRB); Z UMa (SRB); ST UMa (SRB); R UMi (SRB); Y UMi (SRB); SS Vir (SRA); and SW Vir (SRB).

As before, the styles used are those you will find in the relevant document and there are a few cases of duplication where more than one light-curve is given. In most cases I have been able to give the subtype in parentheses after each listed star.

From the foregoing list, the following stars have TA finder charts on the accompanying CD-ROM:

V727 Cas; V386 Cep; V482 Cyg; and SV Sag.

There are TA charts on the CD-ROM for the following additional SR stars:

V720 Cas; V2303 Oph; V376 Vul; EQ Dra; V1070 Cyg; AK Peg;

Referring again to the list of SR stars, the following stars have BAA finder charts on the accompanying CD-ROM:

V UMi; V Aql; V450 Aql; V1293 Aql; UU Aur; RV Boo; RW Boo; V CVn; ST Cam; V465 Cas; mu Cep; RS Cnc; W Cyg; AF Cyg; CH Cyg; U Del; EU Del; MV Del; TX Dra; AH Dra; X Her; SX Her; RX Lep; RY UMa; VW UMa; V UMi; V Aql; RW Boo; RX Boo; RR CrB; SW CrB; AF Cyg; ST Her; g Her; SX Lac; RY Leo; ZUMa; V UMi; SS Vir; and SW Vir.

There are BAAVSS charts on the CD-ROM for the following additional SR stars:

NQ Geminorum (a binary pair containing an SR star); U Camelopardalis (SRB); W Orionis (SRB); UW Herculis (SRB); TT Cygni; and X Cancri.

The following stars may or may not be SR variables and so observing them would be especially valuable:

- EV Aqr: the light-curve is included on the CD-ROM, but there is no finder chart.
- V854 Cas: the light-curve is included on the CD-ROM, as is the TA finder chart.
- W Boo: the light-curve is included on the CD-ROM, but there is no finder chart.
- RY Dra: the light-curve is included on the CD-ROM, but there is no finder chart.
- UX Dra: the light-curve is included on the CD-ROM, but there is no finder chart.

- CK Orionis: there is a BAVSS chart on the CD-ROM.
- RY Draconis: there is a BAAVSS chart on the CD-ROM.
- V336 Vul: there is a TA chart on the CD-ROM.

8.3 A naked-eye hypergiant variable star

Before I leave SR stars, I want to highlight one rather special one. The star ρ Cassiopeia is classed as an SRD by the GCVS Research Group. However, this star is not 'merely' a supergiant – it is a *hypergiant*. Its radius is perhaps about 1000 times that of our Sun and its luminosity is about half a million times that of our Sun. This is why the star appears as bright as it does in our sky despite its distance of 3 kiloparsecs. Its spectral type normally varies in the range of F–K over its rough 320 day period. Meanwhile its visual magnitude normally varies between about 4^m and 5^m but historically it has spanned $4^m.1$ to $6^m.2$.

One of the reasons that I have given this star special attention is that its variations can normally be followed with the naked eye. To that end Figure 8.3(a) shows a photograph of the constellation Cassiopeia and Figure 8.3(b) shows a simple map for identification purposes. I have included just the main stars of the constellation, as well as ρ Cas itself, plus three nearby stars which are useful as comparison stars. Their magnitudes are indicated alongside the stars: '49' actually means a magnitude of $4^m.9$, the decimal point being left out to avoid confusion with the stars on the map. In the same way the magnitudes of the other two comparison stars are $5^m.5$ and $6^m.2$. By comparing the map with the photograph you should be able to identify the stars in question. Note I have also included the rather fainter star which is very close to ρ Cas, as this should help you to identify it.

Although you can theoretically follow ρ Cas through its entire cycle by naked-eye alone if you observe under crystal clear skies, you will be able to follow it in more mediocre conditions – and more particularly you will be able to make more accurate magnitude estimates – if you use a small pair of binoculars or a very small telescope. If your binoculars have big object lenses you will find that stopping each down to 25 mm or 30 mm will bring ρ Cas and the two comparison stars down to a brightness which is more suitable for accurate visual magnitude estimates. The same of course goes if you are using a telescope.

Another reason I have singled out this star is that it has a history of behaving very oddly. To start with its spectrum is very peculiar – with a very complex mix of emission and absorption features. In late 1945 it began a 6 month slide to an unusually deep minimum brightness – at sixth magnitude – which it maintained for nearly a year. At the same time its spectrum changed from type F to type M. Then it began to recover and it returned to nearly its 'normal' brightness and spectrum by mid-1947. In the last few years it has shown peculiar changes to its spectrum and an increased unsteadiness in its already unsteady brightness. From mid-1999 to mid-2000, for instance, the star erratically climbed

Less regular single-star variables

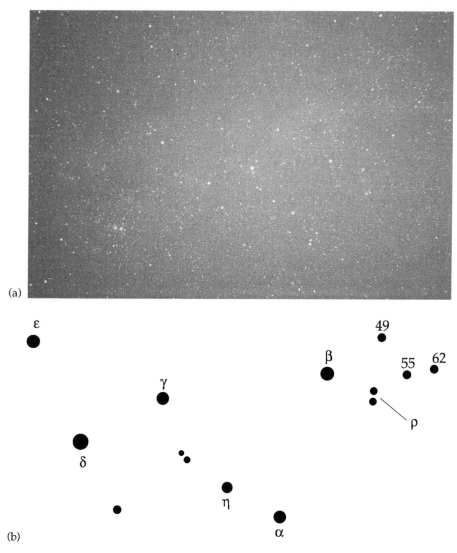

Figure 8.3 (a) The constellation of Cassiopeia, photographed by the author. (b) Key map showing the positions of the variable stars ρ Cas and γ Cas (see text for details).

from $5^m.0$ to $4^m.4$ and then it plunged to $5^m.4$ in about six months, taking about the same time to recover to $4^m.5$. All the time there were marked changes to the star's spectrum. The erratic behaviour continues up to the time I write these words.

Theorists speculate that the unstable outer envelope of this star undergoes episodes where it collapses downwards, like a deflating chewing-gum bubble, onto the inner portions of the star. Amid the chaos there are sure to be powerful shock waves and a violent rebounding of the material. Undoubtedly much material is heaved away into space by the rebound. Thanks to its bloated size and weak surface gravity this star experiences significant mass loss through powerful stellar winds, even without these eruptions.

This star probably weighs in at something around 30 solar masses, maybe a little more. Many astronomers expect this to be the next candidate to go supernova (though probably not until after several tens of thousands more years at the very least). ρ Cas is most definitely a star to be kept under observation.

8.4 L (slow irregular variable); LB and LC stars

This class of variables mostly encompasses those pulsating swollen red giants and supergiants whose variations are very slow (taking years for the star to go from minimum to maximum brightness) and show no signs of periodicity. However, some stars which are initially assigned to this class are subsequently moved into another class (usually SR) when enough observations have been gathered to reveal some periodicity under mathematical scrutiny. There are two subclasses: LB stars are mostly giants and LC stars are supergiants, both spanning a range of spectra from K to S.

The following stars have light-curves on the accompanying CD-ROM:

BR CVn; V770 Cas (LB); V930 Cyg(LB); BZ And (LB); psi 1 Aur(LC); GO Peg (LB); KK Per (LC); PR Per (LC); V770 Cas (LB); AS Cep (LB); and WY Gem (one star of the binary pair is of type LC).

An example light-curve of each subtype is reproduced in Figure 8.4. Of those in the foregoing list, only V770 Cas has a TA chart on the accompanying CD-ROM and only BR CVn is represented by a BAVSS chart. However, you will find charts for the additional LB-type stars GO Pegasi and W Canis Majoris in the BAAVSS chart file. You will find charts for still more L stars in the TA chart file: V451 Cas (LB); UY And (LB); V353 Gem (LB); V1152 Cyg (LB); and NR Vul (LC).

The following stars may or may not be of type L and so are especially worthy of observation:

- V826 Cas: there is both the light-curve and a TA chart for this star on the CD-ROM.
- V451 Cep: there is both the light-curve and a TA chart for this star on the CD-ROM.
- T Cyg: there is a light-curve on the CD-ROM, but no finder chart.
- RU Crt: there is a TA chart for this object on the CD-ROM, but no light-curve.
- NSV 24346: there is a TA chart for this object on the CD-ROM, but no light-curve.
- NSV 12441: there is a TA chart for this object on the CD-ROM, but no light-curve.

Less regular single-star variables

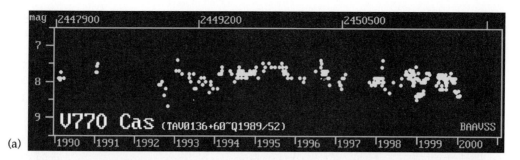

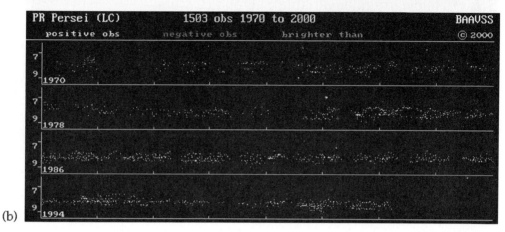

Figure 8.4 (a) The light-curve of V770 Cas (type LB). (b) The light-curve of PR Per (type LC). Courtesy BAAVSS.

8.5 Other pulsating variable stars

Space in this book is pressing, so I must leave it to you to investigate the remaining types of pulsating variable star. A good start can be made by looking up the GCVS entries in the 'Variability types' document on the CD-ROM. However, here follows an extremely brief overview of the remaining types of pulsating star:

ACYG (α *Cygni*) stars are supergiants of spectral classes B_e and A_e. Their periods range from days to weeks but are of small brightness amplitude and so suitable only for precision photometry.

BCEP (β *Cephei*) stars are subgiants of spectral classes O and B with periods of a few hours and small brightness amplitude. They are unsuitable for visual brightness estimation. The BCEPS subtype have periods that are a small fraction of an hour.

DSCT (δ *Scuti*) stars are another batch of short-period and very-small-amplitude stars (spectral types A–F) which are to be avoided until such time as you become a virtuoso of precision photometry.

GDOR (γ *Doradus*) stars are F-type dwarfs with periods of a few hours to a little over a day. Again their tiny amplitudes preclude all but precision photometry.

8.5 Other pulsating variable stars

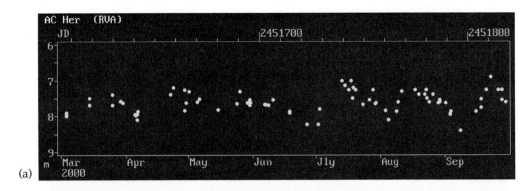

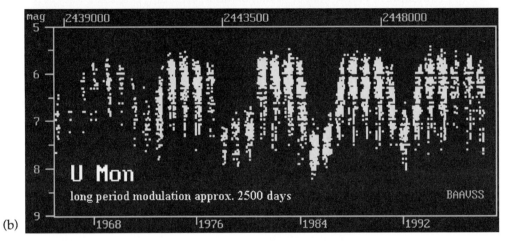

Figure 8.5 (a) The light-curve of AC Her (type RVA). (b) The light-curve of U Mon (type RVB). Courtesy BAAVSS.

- *LBOO* (λ *Boötis*) stars are dwarfs of spectral types A–F. In many ways they are similar to DSCT stars but some of them do have amplitudes large enough for amateur work.
- *PVTEL* (*PV Telescopii*) stars are supergiants whose B-type spectra show unusually strong lines of helium and carbon. Their periods range from a couple of hours to about a day but, yet again, their small amplitudes make them unsuitable for all but precision photometrists.
- *RPHS* is an acronym for *rapid pulsating hot subdwarf* variable stars. With periods of just a few minutes and small brightness amplitudes they are definitely not for the visual observer or even the tyro photometrist.
- *RV* (*RV Tauri*) stars are supergiants whose spectra vary from types F to G at maximum brightness and K to M at minimum brightness. Their brightnesses can change by 3 or 4 magnitudes over periods from about 1–5 months. *RVA* (alternatively *RVa*) stars maintain their average magnitudes from cycle to cycle. *RVB* (alternatively *RVb*) stars have mean magnitudes which slowly change over a span of about 2–5 years, by as much as 2^m. These ancient and bloated

stars are beginning their transformations to white dwarfs. Figure 8.5 shows an example light-curve for a star of each subtype. These are also featured on the CD-ROM, as are light-curves for other SV stars, along with a chart for one of them. These are all listed at the end of this section.

SXPHE (*SX Phoenicis*) stars are population II subdwarfs of spectral types A–F with periods of an hour or two. Some of them have magnitude changes of up to $0^m.7$. They are intriguing because their brightness variations are a blend of several modes of oscillation. However, you will have a hard job in finding any which are bright enough for your equipment.

ZZ (*ZZ Ceti*) stars are actually white dwarfs whose brightnesses vary by tiny amounts in periods of about half an hour. The subtypes, *ZZA*, *ZZB* and *ZZO* have different compositions (and hence different spectra). They are not suitable fare for us.

You will find the light-curves of the following RV stars on the accompanying CD-ROM (the subtype is given in parentheses where known):

LX And; AC Her (RVA); U Mon (RVB); R Sct (RVA); RV Tau (RVB); V Vul (RVA); U Mon (RVB); TX Per and (RVA).

There is a TA chart of R Sge (type RVB) on the CD-ROM.

8.6 RCB (R Coronae Borealis) stars

R Coronae Borealis (*RCB*) variables are supergiant stars with a very wide range of spectral types (B–R). As is normal for such bloated stars, they do pulsate with semi-regular periods of 1 month to several months but their amplitudes tend to be less than $0^m.5$. There is nothing special in that. However, they are remarkable stars because of a particular trick they perform every so often.

As was discovered to be the case for the prototype star in 1795, these stars can hover about their mean brightnesses for several years at a time and then suddenly, and over a span of about a month to several months, their brightnesses drop by many magnitudes. How many magnitudes varies with each fading episode but the drop in brightness can be as much as 9^m. They can recover their brightnesses in just a few months or they can remain unsteadily dimmed for years before they recover. During their dimming and recovery phases their brightnesses also fluctuate erratically on a short-term basis.

Figure 8.6(a) shows the light-curve of R Coronae Borealis as observed by BAAVSS members from 1921 to 2000. Closer looks at the fading episodes of 1995–1996 and 1998–1999 are provided in Figures 8.6(b) and 8.6(c), respectively.

This sort of behaviour marks RCB stars as eruptive variables. In Chapter 6 we encountered other eruptive variables, though the causes of their variabilities are very different to that of the present case. So are their ages. Unlike the pre-main sequence nebular variables described in Chapter 6, RCB stars are extremely old. Their spectra show that they are depleted in hydrogen but rich in helium and carbon. It is quite likely that they have evolved past their planetary

8.6 RCB (R Coronae Borealis) stars

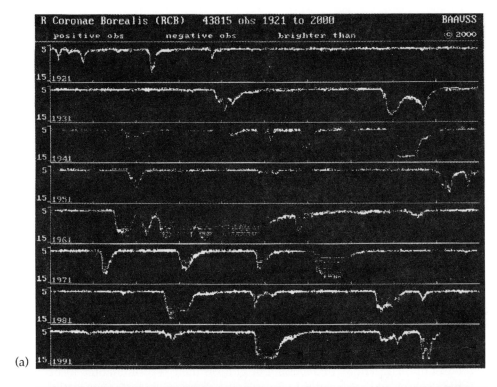

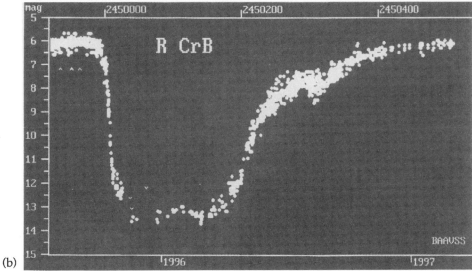

Figure 8.6 (a) The light-curve of R CrB spanning 1921–2000. (b) The light-curve of R CrB during the 1995–1996 fading episode. (c) (overleaf) The light-curve of R CrB during the 1998–1999 fading episode. Courtesy BAAVSS.

Less regular single-star variables

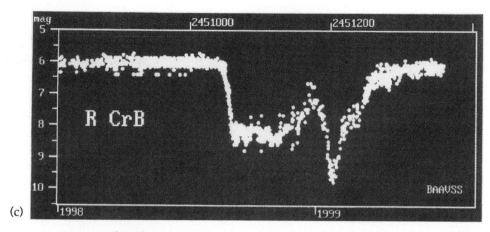

(c)

Figure 8.6 (cont.)

nebular forming stages (spectroscopic traces that might indicate a largely dispersed planetary nebula have been detected in one or two cases) during which they lost much of their hydrogen envelopes. Mind you, there is much diversity of opinion in the astronomical community about the true natures of these stars and how they got to be as they are.

What is the cause of their episodic fadings? Well, one strong clue is that at the onset of their periods of declining brightness their predominantly absorption-type spectra are replaced by a spectrum rich in emission features. With time the photospheric features in the spectrum fade and the emission features also change, then fade. Later, when the brightness is rising the spectrum gradually reverts back to the normal photospheric version.

Opinions vary about the details but it seems fairly certain that the fadings are caused by episodic condensations of solid carbon particles in the upper photospheric, or perhaps in the chromospheric, layers of the star. In effect, every so often the star 'soots up' like a spark plug in an old lawn mower! The sooty layer obscures much of the rest of the radiations from the star. Infrared spectroscopy shows the expected infrared excess that accords with the carbon particles heating up as they absorb the stellar radiations. It is after this smoky cocoon is driven outwards by the radiation and gas pressures from the star that the star can shine forth once again with its normal brilliance.

At the time of writing there are only about three dozen examples of stars known which behave like R Coronae Borealis. However, they are particularly fascinating and well worth observing and so this type of variable star is disproportionately well represented on the accompanying CD-ROM. The following RCB-type stars have light-curves on the CD-ROM:

DY Cen; V854 Cen; WX CrA; R CrB; RZ Nor; RY Sgr; VZ Sgr; GU Sgr; SU Tau; RS Tel; UV Cas; R CrB; V482 Cyg; SV Sge; SU Tau; S Aps; U Aqr; UW Cen; V CrA; WX CrA; R CrB (1); R CrB (2); RZ Nor; SV Sge; V1860 Sgr; SU Tau (1); SU Tau (2); and SU Tau (3).

Of those in the foregoing list, the following stars have TA charts on the accompanying CD-ROM:

V482 Cyg; U Aqr; and SV Sge.

Again from the list of light-curves, just R CrB also has a BAAVSS chart at 'A' and 'B' scales on the accompanying CD-ROM. The 'A' chart covers a span of 18°, which allows for easy identification of the star in the very distinctive bowl shape of the main stars of Coronae Borealis. The 'B' chart covers 9° (very suitable for low-power binoculars) and a complete comparison star sequence. Do bear in mind, though, that at the times of deepest minima R Coronae Borealis does require a moderate-sized telescope for visual detection.

The following stars may or may not genuinely be of RCB-type:

- CG Cam: there are two light-curves on the CD-ROM, but no charts.
- DY Per: there is a light-curve and a TA chart on the CD-ROM.
- Z UMi: there is a light-curve on the CD-ROM, but no chart.
- XX Cam: there is a light-curve and a BAAVSS chart on the CD-ROM. There is very strong evidence to suggest that this star really is of RCB type.
- SY Hya: there is a light-curve on the CD-ROM but no chart.

8.7 GCAS (Gamma Cassiopeia) and B[e] stars

GCAS (or *Gamma Cassiopeia*) stars are named after the prototype γ Cassiopeia. As is shown in Figure 8.3(a) and (b), γ Cassiopeia is the middle of the 'W' formation of stars that form this constellation. Once considered a steadily shining star of magnitude $2^m.3$, it suddenly brightened to $1^m.6$ in 1936. After that episode it remained of variable brilliance, sometimes getting as bright as 2^m but occasionally fading to about $3^m.3$. Up to the time I write these words, its brightness mostly lies in the range $2^m.2$–$2^m.4$. It spectrum is classed as B_e, meaning that it is of this spectral type but with the addition of emission features. This suggests that the star is shedding material.

γ Cassiopeia appears to be very rapidly spinning and it is preferentially throwing off material from its equatorial zones as a result. This material must create shells and rings around the oblate form of the star. This is a star well worth keeping an eye on. As a bonus it is especially easy as there are several naked-eye stars in the immediate vicinity which make excellent comparison stars. I will leave it to you to look up their visual magnitudes.

The GCVS Research Group classify GCAS stars as '. . . rapidly rotating B III–IVe stars with mass outflow from their equatorial zones . . .' Other stars of spectral type B that behave similarly but do not quite fall into the GCAS classification (particularly whose brightness variations appear not to be related to shell-forming events) are simply denoted B[e] stars.

Another naked eye star of GCAS-type is δ Scorpii. Prior to mid-2000 this giant (spectral type B0) star was of magnitude $+2^m.3$. Since then it brightened to

$+1^m.6$ by mid-2002 and is still brighter than normal. Even a simple star atlas will help you locate this star and those around it can act as comparison stars (I will leave it to you to look up their magnitudes). This is another star which is easy to find, fairly easy to observe, and well worth keeping an eye on.

GCAS and B[e] stars are classed as eruptive variables.

8.8 Other single-star eruptive variables

As is the case for Section 8.5, this section contains some 'briefly noted' types of variable star which I will leave it to you to research further. You might make a start by looking up their entries in the 'Variability Types' file on the accompanying CD-ROM. These are all classed as eruptive variables.

SDOR (*S Doradus*) stars. These are highly luminous stars (spectral types B–F with emission features) showing erratic (though occasionally cyclic) brightness changes, often of several magnitudes amplitude. They are often associated with nebulosity, though I should stress that these are not young stars of the nebular variable type. They undergo powerful eruptions which blast away much of the stellar material. Probably the most famous example of this variable type is the far southern star η Carinae.

UV (*UV Ceti*) stars. Of spectral types K–M, their brightnesses can rocket upwards by as much as a few hundred times in less than a minute! You will often find these referred to as *flare stars* in the literature. The variations appear to be caused by an extreme form of the solar flare phenomenon that occurs on our Sun.

WR (*Wolf–Rayet*) stars. These are blue stars with peculiar spectra, dominated by emission features of helium with carbon, oxygen, and nitrogen. Hydrogen is virtually absent. When carbon is particularly dominant they are known as WC stars. WN stars show nitrogen most strongly in their spectra. Observations reveal that they are shedding mass at a prodigious rate.

One idea considered for how they formed involved a companion star stripping away much of the hydrogen envelope, so explaining the deficit of hydrogen in their spectra. The latest thinking, though, is that they are formed by O-type stars of at least 40 solar masses. The sheer brilliance of these stars is enough to literally push away much of their hydrogen envelopes, provided the stars are fairly metal-rich to begin with (it is the metals which the light most effectively pushes against). They are fascinating objects but their tiny brightness amplitudes mean that they are not really good candidates for us to observe.

I (*Irregular*) stars. These are a generally poorly studied collection of stars of ill-defined characteristics that do not readily fall into the other categories. Maybe many of them would do so but for the want of sufficient observation. They are subdivided into *IA* and *IB* types depending on whether they are of spectral types O–A or F–M, respectively.

IS (*Rapid Irregular*) stars. These are a disparate bunch with a number of different likely causes for their variability. Many are probably nebular variables (type

8.9 Rotating variable stars

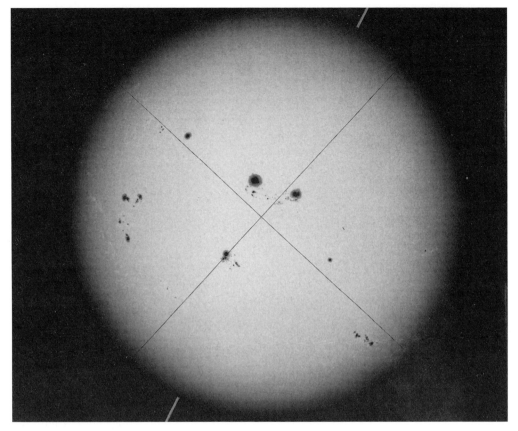

Figure 8.7 Photograph of the Sun taken using the photoheliograph at Herstmonceux on 1970 April 9. Courtesy the Royal Greenwich Observatory.

INS). They are subdivided into types *ISA* and *ISB* according to their spectral types, in the same way as for the irregulars.

8.9 Rotating variable stars

This is another class of stars that I will only briefly mention. They are not really suitable fare for us, particularly because of their tiny brightness amplitudes but also because there are often no examples bright enough for amateur photometry. You will find them, and details about them, listed on the CD-ROM as types: *ACV*; *ACVO*; *BYDRA*; *ELL*; *FKCOM*; *PSR* and *SXARI*.

Just as our Sun has sunspots on its photosphere (see Figure 8.7), so do other stars. Some stars are reckoned to have large areas of their photospheres covered in spots, more so than ever is the case for our Sun. As these stars rotate so the 'spotty' areas move in and out of view from us. Hence the total flux we receive changes a little as a result.

Some other stars may not be so heavily spotted but have irregular brightness distributions and, for those in binary systems, can be spinning so fast on their axes that they are markedly elliptical. As they progress around their binary orbits so the profile they present to us varies – and it is this varying profile that generates the brightness variations we see.

Having mentioned binary stars in connection with rotating variables, in the next chapter we will find plenty more binary systems which produce brightness variations through a variety of other causes. A great many of them are very suitable and especially exciting candidates for us to observe.

Chapter 9
Eclipsing binary stars and novae

Roughly half of the stars in our Galaxy are single. The rest are gravitationally bound into systems of two or more members orbiting around the common centre of gravity. Most often these stars appear as one through a normal telescope unless certain special techniques are used to resolve them. However, the binary nature of the system can cause any of a number of different phenomena to occur which result in visible brightness variations. In many cases these brightness variations are extreme and the phenomena causing them are particularly violent.

If you want a detailed account of a major part of this rather complex subject then I can recommend the excellent book *Cataclysmic Variable Stars – How and Why They Vary* by Coel Hellier, which was published by Springer-Verlag in 2001. The academic level of it is rather higher than in this book and a very large amount of detail is crammed in to it.

Meanwhile, in this chapter and the next I will provide you with just sufficient detail to allow you to prise open a vast new box of stars for your observation, enjoyment, and further study.

9.1 A matter of gravity

Imagine someone using one hand held aloft to twirl a metal bar with a large weight at each end. The person will hold the metal bar in a position that depends on the two weights. If the weights are equal, then the person will hold the bar exactly half way between them. If one of the weights is heavier the person will hold the bar at a position nearer the heavier weight.

This is a good way to visualise the behaviour of two stars orbiting each other. There is no metal bar joining them but the point about which each star moves will always remain on a straight line between the two stars. This point is actually the centre of mass of the system, properly known as the *barycentre*. If the two stars are of equal mass, then the barycentre will be exactly halfway along the

Eclipsing binary stars and novae

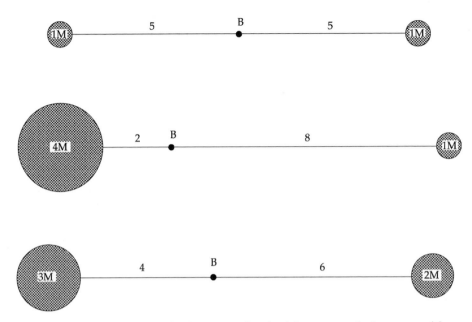

Figure 9.1 Binary star orbits. The distances of each of the stars to the barycentre (B) are in the inverse ratios of their masses in each of these three hypothetical cases.

line joining the two stars. If one star is heavier, the barycentre will lay closer to the more massive star. In fact, the ratio of the distances from the centre of mass of each star to the barycentre is in the inverse ratio of the masses of the stars. Figure 9.1 should make this clear.

Of course, real stars often have elliptical orbits which is where the 'weights at the ends of a metal bar' analogy is inappropriate unless one also allows the bar to lengthen and shorten as it is twirled round.

Isaac Newton was the first to describe mathematically the characteristics of gravity. His 'Universal Law of Gravitation' explained, amongst other things, the behaviour of the planets of our Solar System as they orbit our Sun. This behaviour had been previously summarised by Johann Kepler's empirically derived 'Laws of Planetary Motion'. If you are rusty on such matters, they are described in my book *Astronomy In Depth* (published by Springer-Verlag in 2003). It is from these laws that a simple relationship for stars in a binary system follows:

$$T = \sqrt{\frac{A^3}{(M_1 + M_2)}}$$

where T is the orbital period of the system in years, A is the size of the semi-major axis of the orbit in astronomical units, and M_1 and M_2 are the masses of the individual stars in solar masses (i.e. mass of the Sun = 1). If you are not familiar with the term 'semi-major axis', then think of this as an average distance. *Astronomy In Depth* does go into more detail about these matters. The main point I am making here is that the period of revolution of the stars is determined by the sum of the masses of the stars and the distance separating them. If the total mass in the system is larger, then they move more slowly and the orbital period (the time taken to go once round) is longer. If, though, the stars are closer to each other, they orbit faster and the period is less.

9.2 Eclipsing binary stars

The star β Persei normally shines with a magnitude of $2^m.1$. Every 2.867 days, though, it suddenly fades down to magnitude $3^m.3$. Then it recovers to its full brightness again, the whole episode lasting about 10 hours.

Geminiano Montanari of Bologna is usually credited with first noticing this star's fades in 1669. However, the ancient Arabic scholars called this star 'Ras-Al-Gul', meaning 'the head of the demon' and there has been speculation as to whether the sharp-eyed Arabs had noticed this star's behaviour. Our modern proper name of 'Algol' for β Persei derives from the older Arabic name.

A little over a century after Montanari's observation, John Goodricke established the periodic nature of the light variations and later H. C. Vogel showed spectroscopically (from a cyclic Doppler shifting of the spectral lines) that the star was in orbit about a companion. It seemed obvious that the orbital plane of the star about the unseen companion lay close to the direction in which we view the system. In that way, once every orbit the companion would pass between us and the bright star and block off some of the light in our line-of-sight.

Could the companion be a huge planet? Concrete evidence about its nature had to wait until early in the twentieth century when Joel Stebbins applied accurate photometry to Algol. He found that there was a slight dip in brightness midway between the main fades. This proved that the companion body was itself emitting light, even if it was very much less bright than the main star. So, the companion was also a star, albeit a much fainter one.

Figure 9.2 explains what is going on in the Algol system. When the dim companion star obstructs the light of the primary (position 1 on the diagram) a large dip in brightness is produced (the points shown as 1 on the graph). Half an orbit later the dim star is partially obscured behind the bright star (position 3 on the diagram). Some of the dim star's contribution is then cut off, producing a small dip in the total light from the system that we receive (the points shown as 3 on the graph). At other times we see the main star and the companion star both unobscured (for instance at positions 2 and 4 on the diagram) and receive maximum light from the system (note the corresponding sections of the graphical plot).

I should make clear that Figure 9.2 is inaccurate in that the secondary minimum in the light-curve has here been greatly exaggerated to make it clear. In actuality the secondary dip is a mere $0^m.06$, which is why it took Stebbins' pioneering photoelectric photometry to reveal it.

Prolonged and painstaking study of this eclipsing binary system has revealed further details. For instance, by 1859 Friedrich Argelander had collected enough

Eclipsing binary stars and novae

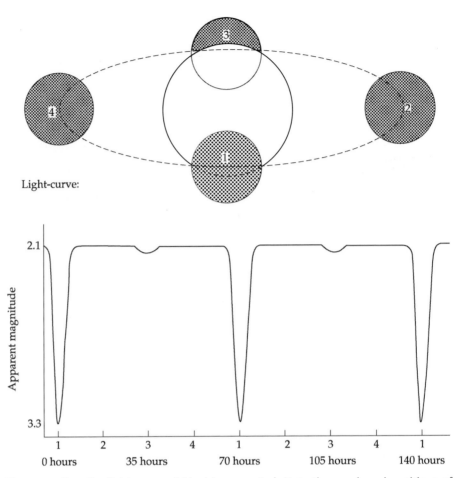

Figure 9.2 How the light-curve of Algol is generated. Note the numbered positions of the companion star in its orbit and how they correspond to the numbered times in the light-curve. The depth of the secondary minimum is greatly exaggerated for the sake of clarity. See text for full details.

observations to be sure of a cyclic, 680 day period, slight variation in the times of minima. This proved the presence of a further massive body. It turns out that the close eclipsing binary system itself orbits another star with this 680 day period. Most of the visible light from the system comes from the main star (Algol A) which is a dwarf of spectral type B.

The close-by companion (usually known as Algol B) is a subgiant of spectral type G. Algol A has a mass 3.6 times that of the Sun, while Algol B has a mass of 2.9 times that of the Sun. The stars orbit the barycentre at distances which are the inverse ratios of their masses. The distance separating the A and B stars is 6.4 astronomical units (960 million km).

Even now we cannot say we have all the facts about the Algol system pinned down. There is still some discrepancy in the drift of the times of minima which

9.2 Eclipsing binary stars

some have speculated means that there must be further companion stars. However, there is some evidence for some matter transferring between the stars. There may even be some magnetic interaction to add to the mix! It is much more likely that these factors have produced the very slight remaining discrepancies.

Before leaving Algol, we should note that there is another peculiarity about the system: the less massive of the two main stars appears to be more evolved than the other one. Yet they should be of the same age (assuming both formed together). We would have expected the more massive star to have run through its evolutionary cycle faster than the other one, yet the reverse appears to be the case. This is another reason for thinking that mass transfer has played an important part in this system. I have more to say about the evolution of interacting stars in the next chapter.

For us to see brightness variations due to the eclipses of any particular binary system the orbital planes of the stars must be very close to our line-of-sight. This situation will occur only infrequently but, since there are so many binary systems on view to us, it does mean that a large number do show this effect. All such stars are denoted as variables of type *E* by the GCVS. You will find them in the 'Variability types' document on the accompanying CD-ROM listed as 'Close Binary Eclipsing Systems'.

Obviously, the way the light varies will depend on many factors peculiar to each system. As we have seen, the masses and separations of the stars will determine their orbital periods. The brightness of each component has a very obvious bearing but so do the shapes of the component stars.

Why shapes, you might ask, surely aren't stars just spherical orbs? Normally yes, but when stars are close together their mutual gravitational attractions can distort each star into an egg shape, with the sharpest ends of the eggs pointing at each other. There are also some stars which are very fast spinners and are oblate in shape as a result of that. The shapes of the stars and the precise angle of the orbital plane to our line-of-sight then has an effect on the light variation while each star is being progressively hidden by the other.

Concerning fast spinning stars, I should mention that there is a tidal breaking effect in operation between close binaries. Some rotational energy is lost as material in a star is pulled up into the bulge. This is an effect which is greatest for the more inflated star of the system. In time this slows the stars down and most often results in the larger of the two stars (and maybe both of them if they are both large) becoming locked into a rotation period which is the same as the orbital period. This is known as *synchronous rotation*. Hence the same part of one (or sometimes both) stars will constantly face the other in the same manner that our Moon now always has the same hemisphere towards the Earth. In certain cases magnetic interactions can also be strong enough to produce synchronous rotation.

Another factor that affects the light-curve is misleadingly called *reflection*. This is where the close-by stars shine on each other to the point that they cause localised heating of the star surfaces. The hotter patch on each star shines more brightly as a result. Add in to the mix that each star can be any of a variety of different sizes and colours and you have a very complicated mix of factors which will determine the light-curve of any given eclipsing system.

The GCVS has defined three tiers of classification for E-type stars. The first is based on the shapes of the light-curves. Thus we have types *EA*, *EB*, and *EW*. These denote *Algol*-type, *Beta Lyrae*-type and *W Ursae Majoris*-type stars (as always, named after the group prototype star in each case). The main difference between these types is the closeness of their orbits and consequent distortion in their shapes. EA stars have the minimum of distortion (Algol A is nearly spherical but Algol B is somewhat distorted) EB stars are significantly ablate, and EW stars are very oblate and touching (or very nearly so).

There is a stand-alone category of *EP* stars for those showing planetary transits. Of course, the amplitudes of the light-curve variations of these are tiny in the extreme.

The second tier of classification is based upon the physical characteristics of the component stars in the system. Thus we have: *GS, PN, RS, WD,* and *WR*. The third tier is based upon the degree to which the stars fill their Roche lobes (Roche lobes are explained in the next section). So, in this tier we have: *AR, D, DS, DW, K, KE, KW,* and *SD*. You will appreciate that a proper account of all of these subtypes of eclipsing binary systems (let alone the various combinations of them) would take a very large amount of space in this book, so please refer to the 'Variability Types' document on the CD-ROM for the definitions for each type of system.

Of course, any particular system may have a place on one, two, or all three of these tiers. Hence we encounter eclipsing systems denoted as, for example, EB/WR or EA/DS/RS, or with a variety of other permutations. It could even be that one of the stars of a system is variable in its own right (for instance it might be a pulsating star) and yet it also takes part in variations due to eclipsing! Such a star might be denoted as LC + E, to take just one example.

The following eclipsing systems have light-curves on the accompanying CD-ROM (with the types given in parentheses – in many cases these have components of types which you have yet to encounter but you will do so later in this chapter or in the next one):

EM Cyg (UGZ + E); V139 Cyg (E + NC); EX Dra (UG + E); U Gem (UGSS + E); AM Her (AM/XRM + E); HZ Her (AM + E); BX Mon (ZAND + E); IP Peg (UGSS + E); LX Ser (EA + UG?); V Sge (E + NL); BF Cyg (ZAND + E); CL Cyg (EA/GS + ZAND); EM Cyg (UGZ + E); U Gem (UGSS + E); eta Gem (SRA + EA); V1413 Aql (ZAND + E); CL Cyg (EA/GS + ZAND); V1329 Cyg (E + NC); and WY Gem (LC + E).

Of these the following examples also have TA charts on the accompanying CD-ROM:

EM Cyg; V139 Cyg; IP Peg; LX Ser; V Sge; V1329 Cyg.

There are TA charts for the following additional stars on the accompanying CD-ROM:

UV Per (EB); UU Cnc (EB/GS); BH Lyn (EA + NL?); AN UMa (E + XRM); V836 Her (EA); LY Del (EA); V1068 Cyg (EA/RS); DL Cyg (EA/DM); and EK Cep (EA/DM).

Referring again to the list of light-curves, BX Mon and BF Cyg also have BAAVSS charts on the accompanying CD-ROM. There are also BAAVSS charts for the following stars:

IP Pegasi (UG + E); RZ Cassiopeiae (EA/SD); EE Pegasi (EA/DM); IQ Persei (EA/DM); HU Tauri (EA/SD?); W Ursae Majoris (EW/KW?); AR Aurigae (EA/DM); EO Aurigae (EA/DM?); LY Aurigae (EB/SD?); RS Canum Venaticorum (EA/AR/RS); U Cephei (EA/SD); SS Ceti (EA/SD); Z Draconis (EA/SD); TW Draconis (EA/SD); RW Geminorum (EA/SD?); beta Lyrae (EB); ST Persei (EA/SD); and Z Persei (EA/SD).

If you are interested in examples of pure eclipsing binaries that you can follow with your unaided eyes, then I recommend Algol (β Persei) and β Lyrae. They can easily be located using a simple star map and I will leave you to select suitable comparison stars (though there is a chart of β Lyrae in the BAAVSS charts file on the accompanying CD-ROM for you to use if you wish). They are good stars to practise on. Do bear in mind, though, that it is the more complex and interacting systems which are most worthy of attention.

9.3 Introduction to interacting stars

The stars which we are presently interested in orbit rather closely to each other. As a consequence their mutual gravity can do more than just keep them in their orbits. Their shapes can be distorted and matter can even be passed from one star to the other. Knowing how their gravitational fields interact is important in understanding their behaviour in these circumstances.

Take a small object from the floor and place it on a desk. Or take it from the desk and place it on a high shelf. You have increased the *gravitational potential energy* of that object because you have moved it a little further away from the centre of the Earth. Each level – say the floor, the table-top, and the high shelf – corresponds to a given value of *gravitational potential*, which is the energy possessed by every kilogram of matter in any object placed there. So, the potential energy gain of a brick as a builder lifts it from the ground to the top of a wall is equal to the difference in the gravitational potentials between the ground level and that at the top of the wall multiplied by the mass of the brick.

In the case of the small object you moved, the energy gained by it was provided by you because you (via your muscles) used some of your energy to lift the object against gravity. The same was true for the builder lifting the brick.

Eclipsing binary stars and novae

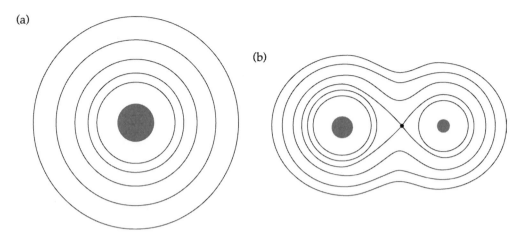

Figure 9.3 (a) The equipotentials surrounding an isolated star are spherical in shape. (b) The equipotentials of the two stars in a binary system are mutually distorted.

The levels of common gravitational potential form *gravitational equipotential surfaces*. On the small scale you would describe these as imaginary horizontal flat sheets, each situated at its own height. You would also describe the surface of the Earth as flat at the same scale. On the larger scale, gravitational equipotentials form spheres which surround our spheroidal Earth. Any single massive object, for instance a star, will be surrounded by imaginary concentric spheres which represent these gravitational equipotential surfaces.

Figure 9.3(a) represents a two-dimensional cross-section through the star and the gravitational equipotential surfaces. Of course, the diagram consists of a series of concentric circles, in the same manner as would an onion sliced in half, though the layers really form spheres.

Remember, these gravitational equipotential surfaces define planes on which any object would have common values of gravitational potential energy. Bring another massive object up close to the first one and the gravitational equipotential surfaces are no longer spheres. The gravitational fields of each of the objects causes a mutual distortion of the gravitational equipotential surfaces. A pair of closely orbiting stars would have an arrangement of equipotentials like that represented in Figure 9.3(b).

The surface of a fluid body will actually lay on a gravitational equipotential surface. So, a body placed anywhere on it will have a common value of gravitational potential energy. If some transient phenomenon were to cause part of the surface of the body to heap into a 'mountain' then, when the phenomenon was over, gravity would act to make the whole surface of the body conform to the original equipotential surface once again.

Stars are sufficiently fluid in nature to conform to their equipotentials. So, stars in binary systems become deformed because of the distortion of their equipotentials. This deformation is least when the masses of the stars are small and/or

9.3 Introduction to interacting stars

Common gravitational equipotential
surface enclosing Roche lobes

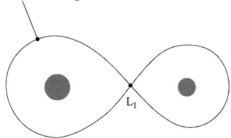

Figure 9.4 At one particular common value of gravitational potential the equipotential surfaces meet at the L_1 point. The spaces enclosed by each of these surfaces around each star are known as the Roche lobes of the stars.

the separation of the stars is large (such as is the case for EA-type binaries). It becomes large when the opposite conditions apply (in EW-type binaries, for example).

There is a given value of gravitational potential in any binary system which corresponds to a special case of the equipotential surfaces of that value around each star actually meeting. The term *Roche lobe* is used for the space enclosed inside that common equipotential surface for each star (see Figure 9.4). The location in space where the Roche lobes meet is of major importance and is known as a *Langrangian point*. More properly this is known as the *inner Langrangian point* or *L_1 point* because there are several others at various positions in space around the system. The others need not concern us.

The L_1 point is a point of unstable equilibrium between the gravitational fields of the two stars. An object placed at this point would experience equal gravitational pulls from each star. Theoretically the object could remain poised there. However, give the object a small nudge in the direction of one of the stars, then the attraction of that star would progressively dominate and the object would accelerate towards it. If the object was nudged from the L_1 point towards the other star, then that star would dominate and the object would fall towards it instead.

If you wanted to transfer an object from one star to the next, then the most energy-efficient way you could do it would be to launch the object from the first star, send it through the L_1 point and let it fall towards the other star. All other routes would require a lot more energy.

If for any reason some matter is thrown off one star of a binary pair it will most easily cross from one star to the next if it passes through the L_1 point. All that is required is that the matter is endowed with sufficient energy to lift it from the photosphere of the star to the L_1 point. In quantitative terms, the minimum energy required per kilogram of matter is equal to the difference in gravitational potentials of the photosphere and the Roche lobe.

Eclipsing binary stars and novae

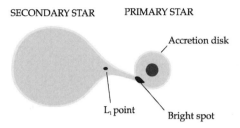

Figure 9.5 This not-to-scale diagram figuratively illustrates matter streaming from the bloated secondary star towards the primary star, where it goes into orbit to form an accretion disk. The point of contact between the matter stream and the accretion disk forms a superhot region, known as the 'bright spot'.

Actually, not only *can* matter flow from one star to its binary partner via the L_1 point, if it has sufficient energy – it *will* do so. Moreover, if the 'donor' star is expanded in size (for instance it might have evolved into a red giant) the energy required to lift matter from its photosphere to its Roche lobe is less and matter can flow from it to its companion even more easily.

In the extreme case a star might have puffed itself up to such a size that it fills its own Roche lobe. In that instance the surface of the star would already be at the L_1 point and matter can pour from the 'donor' star into the Roche lobe of the companion as easily as water runs over a waterfall. This is especially so, as the particles that make up the gas/plasma will themselves have large kinetic energies by virtue of their high temperature.

Such a binary system, where one star fills its Roche lobe and spills material through the L_1 point onto the other star, is known as a *semi-detached binary*. The bloated donor star is referred to as the *secondary* and the recipient star is known as the *primary*. If this seems the wrong way round, it is because the main phenomena causing the apparent variability we see from Earth happen at the recipient star.

A binary system with both stars contained well inside their Roche lobes is known as a *detached binary*. At the other extreme, some binaries have both stars expanded to the point that each fills, or overfills, its Roche lobe. In that case the outer parts of the two stars merge to form a *contact binary*.

In the Algol system, Algol B is a subgiant which has almost, or perhaps only just, filled its Roche lobe. So, it lies on the dividing line between detached and semi-detached binaries. This is also why it shows some slight evidence of mass transfer between the two stars and probably means that it should be classed as a semi-detached binary system.

The matter passing through the L_1 point of a semi-detached binary will spill straight onto the primary star if it is large. If the primary star is compact then most of the matter will tend to go into orbit around it, forming an *accretion disk*. This is illustrated in Figure 9.5. In time the matter in the accretion disk will lose kinetic energy due to interparticle collisions and will finally land on the star. Of course, all the while fresh material may be overflowing from the secondary star to replenish the material lost to the accretion disk.

Where the matter stream hits the accretion disk a 'hot spot' will be produced (astronomers usually refer to this as the *bright spot*). In fact, this can be a very hot spot indeed. In some circumstances the temperature here may soar to several millions of degrees. After all, the energy of the matter from the primary star gets a huge boost as it falls from the L_1 point through the gravitational field of the primary star; especially so if the primary star is reasonably massive but extremely compact – as is certainly the case if it is a white dwarf. It is easy to imagine this stream of matter slamming into the accretion disk, its kinetic energy being converted mostly into heat energy, the rest generating shock waves and ferocious turbulence in the accretion disk.

Such transfers of matter from one star to the other in a binary pair lead to some of the most powerful phenomena we will encounter in our variable star researches. I enjoyed a particularly exciting night as a teenager in 1975, thanks to a particular instance of matter transfer between two stars.

9.4 N, NA and NB (classical novae), and NC stars

As twilight fell on the evening of 1975 August 29 I happened to step outside and glance upwards. A few of the brightest stars were becoming visible against the deep-blue sky. Arcturus and Vega were prominent and I could see several of the brightest stars in Cygnus – but something struck me as odd. Cygnus did not look quite right. Then I realised Cygnus had gained a bright star to the north-east of Deneb! I could hardly believe it – what had I found? Was it a distant supernova? Maybe it was a less powerful, and so closer, *nova* – an erupting star. I telephoned an amateur astronomer friend and he confirmed the new star.

Next, I telephoned the Royal Greenwich Observatory, then at their former Sussex home of Herstmonceux. An astronomer there told me that they were about to turn the 2.5 m Isaac Newton Telescope (which has since been rebuilt and moved to La Palma) onto the object. It was indeed a nova and had been discovered several hours earlier, before darkness fell over Europe, by several Japanese amateur astronomers. I was one of hundreds to independently find the nova that first night.

The original star is not shown on pre-discovery photographs and so was definitely fainter than the twentieth magnitude. When I saw the object it was of the second magnitude and it reached maximum brightness on the following night, attaining magnitude $1^m.8$. Hence the original star had brightened by a factor of several tens of millions in a time period of only a day or two!

Subsequent nights showed the nova in rapid decline. Within a fortnight it had become too faint to see with the naked eye and Cygnus had once more resumed its normal appearance. Figure 9.6 shows the light-curve of this object – *Nova Cygni 1975*, otherwise known as V1500 Cygni.

Eclipsing binary stars and novae

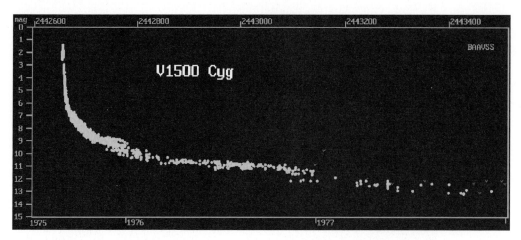

Figure 9.6 Light-curve of the very fast and brilliant nova V1500 Cygni. Courtesy BAAVSS.

About thirty novae appear in our Galaxy each year, though they rarely become bright enough to be visible to the naked eye. The normal configuration producing a nova is that of a highly evolved star of similar mass to our Sun having become a red giant or subgiant star and feeding hydrogen gas onto a white dwarf star via an accretion disk.

As the hydrogen eventually migrates onto the surface of the white dwarf the intense gravity there compresses it. Much of it is compressed to the point that it becomes degenerate, in common with the rest of the white dwarf. The temperature of the hydrogen, already raised by the energy liberated by its fall from the L_1 point to the accretion disk, soars higher and higher as more piles down from the accretion disk.

Actually, the fact that the hydrogen layer has become degenerate exacerbates the increase in temperature. This is because degeneracy pressure tends to resist further increases in compression as more matter piles in on top. Instead the electrons are forced to take up ever higher energy levels and this equates to a higher temperature.

Eventually the temperature of the hydrogen increases to the point that nuclear fusion reactions start. This further boosts the temperature. Degenerate matter does not expand as its temperature increases (this would moderate the nuclear reaction rate if it did) but the rate of nuclear reactions does greatly increase.

The increased reaction rate further boosts the temperature, which further increases the reaction rate, so boosting the temperature even higher. The reaction rate quickly becomes runaway. Reactions that usually only occur deep within a normal star, and which are contained by the body of the star, are suddenly kindled at the surface of the white dwarf. The result is a sudden tremendous release of energy. The gas pressure soars until it exceeds the degeneracy pressure. Suddenly matter is released from degeneracy and blasted into space.

Figure 9.7 Novae Vulpeculae 1979. (a) The nova on 1979 July 20. (b) The nova on 1972 August 20, when it had dimmed considerably. Photographs taken with the 0.33 m astrographic refractor, formerly of the Royal Greenwich Observatory at Herstmonceux. Courtesy the Royal Greenwich Observatory.

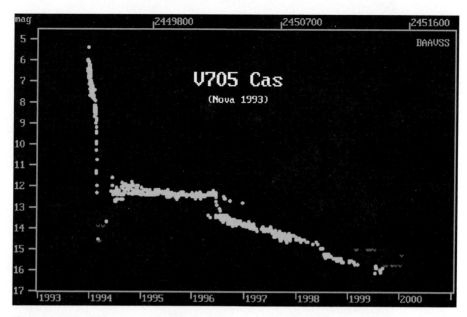

Figure 9.8 The light-curve of V705 Cassiopeiae. The steep dip in brightness just after maximum light is due to the formation of condensed matter in the expanding nebulosity in the aftermath of this nova. However, the obscuration quickly cleared and the brightness recovered, after which the brightness continued its slower decline. Courtesy BAAVSS.

The loss of mass in a nova outburst usually amounts to about one ten-thousandth of the mass of the initial star. The sites of many novae show rapidly expanding nebulosity after the outburst. Some of the energy of the eruption comes from radioactive decay of the daughter products of the nuclear reaction. At maximum, rare examples of novae can shine with an intrinsic luminosity up to 40 million times greater than our Sun, often taking only a few days to reach this peak. More usually novae surge in brightness to between a thousand and a million times brighter than our Sun. Nova Cygni 1975 was notable as one of the brighter examples as well as for its very rapid rise to maximum brightness and its subsequent rapid fall. Others are slower, like Nova Vulpeculae 1979 shown in Figure 9.7.

Interestingly, the formation of a nebula around the nova can often have a marked imprint on the light-curve. Figure 9.8 shows the light-curve of one star, V705 Cas which went nova in 1993. Notice the plunge in brilliance from maximum light to about fifteenth magnitude and the subsequent sharp recovery to twelfth magnitude before the further much slower decline. The decrease in brilliance happened because of particles of solid matter (mostly carbon and silicon) condensing in the nebula as it expanded and cooled. This blocked much of the light from the star, much as smoke can obscure a fire. The subsequent dissipation of this 'smoke' into space let the radiation through once more. Another nova light-curve on the accompanying CD-ROM which also shows this effect well is that of V1419 Aql.

Novae that quickly rise to maximum and fall at least 3^m in 100 or fewer days are denoted as 'fast' and given the classification *NA*. Broadly, slower examples are termed 'slow' and given the classification *NB*. Types NA and NB are often referred to as *classical novae*.

There is another category, *NC*, which is used to denote *very slow novae*. These can stay close to maximum light for years, or even decades. They are sometimes referred to as *symbiotic stars*, the older 'catch all' term for binary stars where mass transfer produces long-term elevations in the brightness of the system but this now tends to be used for systems where the accreting star is not actually a white dwarf but is something less condensed such as a star which is still on the main sequence.

Just how fast and vigorous a nova outburst is depends very much on the mass of the accreting white dwarf. The more powerful surface gravity of a massive white dwarf leads to a much larger explosive release of energy. This is because a much greater pressure is required to release the matter from degeneracy – and so there is more energy to be suddenly released.

9.5 NR stars (recurrent novae)

If matter is constantly streaming from the bloated star to the white dwarf in a close binary system you might wonder whether after one nova explosion yet more material can accrete and then be delivered onto the white dwarf to one day repeat the explosion. The answer to that is yes. Novae can and do give repeat performances. The brighter novae are thought to erupt once in every few tens of thousands of years, on average.

This brings us to another category: *recurrent novae*, denoted *NR* stars. In actuality almost all novae are really recurrent novae (dependent on the subsequent evolution of the system), though this term is only applied to stars where nova outbursts *have been observed to repeat*.

How frequently a nova eruption occurs depends on the mass of the white dwarf and the rate of mass transfer from the secondary star. The mass of the white dwarf matters because for the more massive examples (up to the limiting mass of these objects, which is 1.4 times the mass of our Sun) less matter needs to be collected on its surface before sufficient is compressed to a high enough pressure and temperature to 'go bang'.

The rate of transfer of matter is important because this determines the time taken to accumulate the critical mass on the white dwarf, whatever that value of critical mass might be (a few thousandths of a solar mass for white dwarfs of about 0.6 solar mass to a few tens of millionths of a solar mass for those near the 1.4 solar-mass limit). Systems like T Pyxis show the most extreme form of this type of behaviour because they have the most massive white dwarfs and consequently erupt every hundred years or so.

Other recurrent novae differ in that their component stars are much further apart and so have longer orbital periods. The additional fact that some of them can erupt much more frequently (eruptions occurring every few decades) speaks of modifications and complications to the foregoing simple scenario. Our picture of what is going on in these cases still needs to be finalised.

The star T Coronae Borealis is one example of novae intermediate in terms of brightness and period between outbursts. It is normally of the tenth visual magnitude, but rose to the second magnitude for a brief spell in 1866 and again in 1946. This system is formed from a giant star of spectral type M and a white dwarf orbiting each other with a period of 227 days.

9.6 Novae on the accompanying CD-ROM

You will find light-curves of the following novae on the accompanying CD-ROM in 'BAAVSS lightcurves', with the subtypes indicated in parentheses where known:

V1493 Aql (NA); V1494 Aql; Q Cyg (NA); RS Oph (NR); HR Del (NB); RS Oph (NR); GK Per (NA + XP); V1370 Aql; V1419 Aql; V1425 Aql; V705 Cas(1); V705 Cas(2); V723 Cas; T CrB (NR); V1500 Cyg (NA); V1819 Cyg; V1974 Cyg; Q Cyg; V533 Her; V827 Her (NA); V838 Her; Nova 1998 Mus; Nova 1998 Oph; GK Per(1); GK Per(2); HS Sge; V4361 Sgr; V4444 Sgr; V4633 Sgr; NQ Vul; and PW Vul.

Of those in the foregoing list the following novae have TA charts on the accompanying CD-ROM:

V1493 Aql; RS Oph; V1370 Aql; V1419 Aql; V1425 Aql; V705 Cas; V723 Cas; V1819 Cyg; V1974 Cyg; Q Cyg; V827 Her; V838 Her; GK Per; V444 Sct; HS Sge; and PW Vul.

There is also the BAAVSS chart of AG Pegasi (NC), as well as TA charts of the following additional novae on the accompanying CD-ROM:

V400 Per; QZ Aur (NA); KT Mon; DM Gem; V351 Pup; T Pyx (NR); BY Cir; V992 Sco (NA); V2110 Oph (NC); V977 Sco; V745 Sco (NR); V4135 Sgr; V4742 Sgr; V4171 Sgr; V463 Sct; V4077 Sgr; V603 Aql (NA); V443 Sct (NA); V1460 Sgr; HR Lyr; V4743 Sgr; QV Vul (NA); V1548 Aql; V1378 Aql; V1301 Aql (NA); V368 Aql (NA); V2274 Cyg; V404 Cyg; QU Vul (NA); V1330 Cyg; V2275 Cyg; and OS And.

The following is a list of TA charts for those stars which have undergone nova-like eruptions but whose provenance remains uncertain:

LS And; PQ And; SV Ari; CG Cma; U Leo; V1172 Sgr; V3645 Sgr (NR?); NSV 24587; EU Sct (NR?); and FS Sct (NA?).

There is also a selection of images of novae 'caught in the act' on the CD-ROM. To find these start the CD-ROM and open the 'Index' folder as you would normally do. Under the heading 'Miscellaneous' you will see the link 'Various images'. Click on that and you will see a selection of thumbnails. Clicking on any chosen one of these will deliver a full-size version onto your screen. There are 49 images for your perusal. Of them the following ones are of classical and recurrent novae (listed here in the order you will find them, going left to right along the rows and then row-by-row down the screen):

GK Per (imaged on 1999 February 22 – the famous recurrent nova first seen to erupt in 1901); Nova Aql 1995 (imaged on 1995 May 28); Nova Aql 1999 (on 1999 July 14); Nova Cas 1995 (on 1995 August 27); Nova Cas 1995 (on 1995 September 2); Nova Cygni 2001 (on 2001 August 20); V2274 Cyg (on 2001 August 20); Nova Sgr 2000 No.2 (on 2002 September 26); Nova Sgr 2002 No.3 (on 2002 September 23); V1494 Aquilae (on 1999 December 3); V1548 Aql (on 2001 May 31); V1548 Aql (on 2001 August 20); V1974 Cyg (on 1997 September 2); and XTE1118+480 (an X-ray nova imaged on 2000 March 30).

As ever, in the foregoing list I have used the styles given in each entry in order to help you (and your web browser) find them easily.

9.7 NL stars (nova-like variables)

The GCVS defines the class of objects known as *nova-like variables* (denoted as *NL* stars) as: '... insufficiently studied objects resembling novae by the characteristics of their light changes or by spectral features. This type includes, in addition to variables showing nova-like outbursts, objects with no bursts observed; the spectra of nova-like variables resemble those of novae, and small light changes resemble those typical for old novae at minimum light. However, quite often a detailed investigation makes it possible to reclassify some representatives of this highly inhomogeneous group of objects into other types'.

You should be aware that some other authorities prefer to classify nova-like variables as all stars with sufficiently high mass-transfer rates to maintain the potential for nova-like outbursts.

As a class they are not the easiest subjects to observe. As far as I am aware the brightest of them is IX Velorum (at declination $-49°$ far too far south for all but antipodean observers). Even this object is usually to be found circa tenth magnitude. Most also have rather small amplitudes, so they are subjects for photometry rather than visual observation. Of course, the main point of interest for amateur astronomers is trying to spot early signs of eruptive behaviour if they are true novae or to determine their proper classification if they are not. While I cannot recommend these objects to beginners, some examples of them are featured on the CD-ROM and so I will give them here as an introduction to this intriguing class which you might follow up at some stage in your observing career.

Examples of the light-curves of NL stars (with varying degrees of uncertainty in their classification) on the accompanying CD-ROM are:

V751 Cyg; MV Lyr; V426 Oph; V2204 Oph; V Sge (classified as E + NL); EI UMa; V751 Cyg; SX LMi; and MV Lyr.

Of these, the following have TA charts on the accompanying CD-ROM: EI UMa; SX LMi; V2204 Oph; MV Lyr; V Sge; and V751 Cyg.

There are also TA charts for the following additional objects tentatively classified as NL stars:

BH Lyn (classified as EA + NL); GM Sgr; and V605 Aql.

9.8 Nova hunting

This is a book about observing already known objects with a view to following their brightness changes, not a book about hunting down new examples. However, you are certainly not alone if you hanker to discover new objects yourself – but I must offer a warning. Increasingly professional programs of automated sky-searching are mopping up the new objects to be seen in the heavens. In addition a few highly motivated (and usually very expensively equipped) amateurs are vigorously doing the same. Discovering a genuinely new comet, asteroid, variable star, nova, or supernova from your own backyard is now very much harder to do than it used to be despite modern technology. It is now only fractionally short of 100 per cent certain that someone else will have seen and recorded that new object before you.

If you really are determined to become a nova hunter, then to have even the slightest chance of success you must devote yourself to that task almost entirely. You will have precious little time or energy left for other lines of observational astronomy. If you choose that route, then this is not the book to help you. I do cover nova hunting amongst other subjects in my book *Advanced Amateur Astronomy*, which was published in its second edition by Cambridge University Press in 1997. The one caveat I would warn you about is that in 1996 (when I wrote that edition) it was still reasonable for hunters to use entirely visual means and a pair of good binoculars to do their hunting. While the chances of success will never quite fall to zero using that method, they cannot be very much above that by now.

The way ahead for you is to use photography. This can be by 'old-fashioned' film photography, which is extensively covered in *Advanced Amateur Astronomy*. A 35 mm format camera with a 135 mm telephoto lens, in order to image about $10° \times 15°$ of sky, would be ideal. The camera, loaded with a fast (circa 400 ISO) film, is tracked on the stars for several minutes in order to record faint stars. The modern alternative is to use electronic imaging. This is also covered in the same book, though I have to say that things have greatly moved on since 1996, particularly with the use of digital cameras and webcams.

Photographs (either negatives or electronic images) of the same region of sky are compared with one another by *blinking* – first viewing one, then the other superimposed in position with the first. Anything that changes will show up by suddenly appearing and disappearing as the images are changed. I go into the

9.8 Nova hunting

procedures and equipment in *Advanced Amateur Astronomy*. Using electronic images (perhaps from scanned negatives, though more likely directly from a CCD-based camera) in a computer with the appropriate software to do the blinking on a monitor is the modern way forward.

I do not undertake nova hunting. However, I am always keen to observe these 'things that go bang in the night' when news comes through of new discoveries. If you belong to an observing group which is 'plugged in' to hot news, then so can you. Undoubtedly the most 'plugged in' group of all is TA, The Astronomer Magazine. I give contact details for this organisation in Chapter 1.

Chapter 10
Cataclysmic and symbiotic systems

The various subtypes of novae we encountered in the last chapter are all examples of 'explosive and nova-like variables', or *cataclysmic variables*. In this chapter we discuss some others, particularly the stars known as dwarf novae, which erupt in a different way to novae. The examples of cataclysmic variable we consider in this chapter exhibit less powerful (but often more frequent) eruptions than novae – but those we encounter in the next chapter are those that make the biggest bangs of all.

Almost all of these cataclysmic variables do what they do because of matter transfer between closely orbiting stars. Amongst them are interacting systems which are less aggressive in their behaviour. These are termed *symbiotic stars*. Having mentioned these in the last chapter, I discuss them more fully near the end of this chapter.

First, though, we consider how it is that a cataclysmic variable can come about.

10.1 How to make a cataclysmic variable

Binary star systems are not born as cataclysmic variables. Some never become them but others undergo an amazing transformation in an incredibly short period of time to become one or other of the various types of cataclysmic variable. The transformation involves a number of steps and a variety of processes that lead to the final result. In order to present this rather involved sequence clearly, I here offer my 'thirteen-point quick guide' to the creation of a 'typical' cataclysmic variable system:

1. The progenitor of a cataclysmic variable is a conventional binary star system, consisting of two conventional main sequence stars – one a little more massive than our Sun, one a little less so – orbiting one another with a

10.1 How to make a cataclysmic variable

   separation of the order of a 100 million kilometres and an orbital period of a few years.
2. The more massive star evolves more rapidly than the other (I discussed stellar evolution in some detail in Chapter 6) and eventually reaches the end of its own main sequence hydrogen-fusing stage. It then suffers the usual core contraction and hydrogen envelope expansion to become a red giant. Meanwhile the other star is happily continuing to live its main sequence existence.
3. The red giant star swells to the point that it begins to overflow its Roche lobe and so starts transferring matter via the inner Langrangian point to the other star.
4. However, the barycentre of the system lies close to the more massive red giant star, so this transfer of matter seriously disturbs the distribution of angular momentum in the system. The matter stream robs the system of angular momentum as the matter moves further away from the barycentre in its progress towards the low-mass star. Consequently the orbital separation of the two stars decreases.
5. As the stars creep closer together so the equipotential surfaces (see Section 9.1) crowd together – and so the Roche lobes also contract.
6. The shrinkage of the Roche lobe around the red giant star results in it overflowing the Roche lobe ever more vigorously.
7. The increased rate of matter transfer robs the orbiting stars of more angular momentum at an ever increasing rate and so they move even closer together. This further crowds the equipotentials, further reducing the size of the Roche lobes and increasing the rate of overflowing of the red giant star even more. I am sure that you can see that this is a vicious circle. The mass transfer becomes runaway. In much less than a century after the matter transfer properly got going the former red giant has its entire hydrogen envelope stripped away!
8. Most of the previously massive red giant star is heaped onto the other star. While this was happening the barycentre shifted away from the former red giant and towards the other star, which became the more massive of the two. Meanwhile the helium-rich core of the former red giant is finally rendered naked. This core is, of course, largely electron degenerate (see Section 6.6). *The former red giant has been transformed into a white dwarf.*
9. As I indicated, *most* of the matter from the hydrogen envelope of the former red giant is heaped onto the other star – but not all. Some of it is spilled into the volume of space immediately surrounding and enveloping the two stars. Astrophysicists call this the *common envelope phase*.
10. There is a considerable drag on the two stars as they are do their orbital dance inside their shared shroud of star-matter. Consequently even more angular momentum is extracted from the orbiting stars and their separation shrinks even faster than before. Indeed, in 1000 years or so their separation

shrinks from around a 100 million kilometres, or a little more, to around a million kilometres. So in the end their separation is not much greater than the radius of our own Sun. A consequence of orbiting more closely is orbiting more rapidly. The initial orbital period for the former two main sequence stars was a few years. Now it has become a few *hours*. It is staggering to think of such a vast change happening to a pair of co-orbiting stars in a timescale as short as 1000 years.

11. The angular momentum lost to the orbiting stars is taken up by the shared gaseous envelope. Consequently it expands outwards. This is exacerbated by what I like to think of as a 'cosmic lawn-sprinkler' effect but which is formally known as the *propeller effect*. This is where the gas in the orbital plane of the system is dragged along by the gravities of the stars as they pass by and is spun-up to higher values of angular momentum and kinetic energy. So, the expansion of the gas away from the system happens preferentially in the orbital plane of the stars.

12. The envelope of matter escapes the gravitational influence of the stars to go on expanding into space to form a planetary nebula. This is the alternative way a planetary nebula can form. I discussed the other scenario in Section 6.6. After a few thousand years the planetary nebula disperses enough to render it undetectable from the Earth (though it may be replaced by future versions created by eruptions from the system).

13. The relatively stable system will now include a white dwarf star and will be a cataclysmic system if the masses of the two stars and their separations results in matter transfer. This time, though, the transfer of matter is in the opposite direction to that before. The former red giant lost much matter to the other star and became a white dwarf in the process. Now it is the white dwarf which is on the receiving end of the matter stream. In fact, it is getting some of its own lost material back again!

As I have said, this is a fairly typical scenario as best we understand it. Clearly the masses of the initial stars and their separations will have a significant effect on the nature of the system formed, as will the various natural and enforced evolutionary changes to the stars arising from their interactions. For instance, the primary star might become even more compact than a white dwarf. It might become a neutron star or even a black hole. In fact there are so many ingredients and factors that will affect the outcome that each cataclysmic variable is in its own way unique, despite our best efforts to put them into categories.

As to their final fate, millions or perhaps billions of years in the future, again each system will be unique. However, we can speculate about a typical system. We think that the mass transfer from the secondary star will continue at a reducing rate until a lot of its mass (the rest is ejected from the system by nova-like eruptions) has been returned to the primary. The secondary star will eventually be reduced to a small body with a mass of a few times the planet Jupiter. While this is happening the orbital period is lengthening and the secondary star is

10.2 UG (U Geminorum) stars, aka dwarf novae

retreating from the primary. Finally the mass transfer ceases and the remnant of the secondary star cools and takes on the status of a rather exotic gas-giant planet now orbiting its white dwarf parent star.

10.2 UG (U Geminorum) stars, aka dwarf novae

UG (after the prototype *U Geminorum*) stars show nova-like outbursts which are much less powerful than those of true novae but happen much more frequently. In this respect they are similar to recurrent novae, though they do have other characteristics which justify them having a class of their own. Moreover, we now know that the mechanism for the eruptions is totally different to that of true novae. Even the location of the eruption site is different.

Typically a UG-type star might increase its light output by about 10–250 times. A small number exceed this increase in brightness, while others do not reach 10 times the quiescent value. The decline back to normality takes from a few days to a month or two. The stars showing the greatest brightness jumps take the longest time to decline.

There are cycle-to-cycle variations in the brightness of eruption and the period between eruptions for each system, though there are fairly well-defined averages of these quantities for each example. It is also true that the periods between eruptions are on average longer for those that produce the greatest (again, average value) brightness jumps during outbursts.

At minimum light these stars vary erratically in brightness with amplitudes of $0^m.05$–$0^m.5$ in timescales of seconds, minutes, and hours. We call this effect *flickering*. After a further period of a few weeks to a few months, depending on the system, the star erupts once again.

So, on the surface these stars appear to show a watered down version of a nova's behaviour, albeit repeating outbursts very much more frequently than a 'full-blown' nova. Consequently they are commonly also known as *dwarf novae*, which is often abbreviated to *DN*. You might also come across them referred to as 'SS Cygni stars', named after a particularly bright member of this class. However, you should bear in mind that it is U Geminorum which is the 'prototype star' and SS Cygni is now most usually regarded as one of the three subtypes of UG stars. More about those subtypes shortly.

SS Cygni's apparent visual magnitude is about $12^m.4$ at minimum but it flares to around $8^m.2$ at outburst every 50 days or so. U Geminorum itself does almost as well, normally being around magnitude $14^m.9$ but flaring as bright as SS Cygni does every 103 days, on average.

It was when U Geminorum was in outburst that it caught the eye of John Russell Hind in December 1855. He was actually hunting for minor planets but came across a new ninth magnitude star-like object in Gemini. He expected it to shift its position over the next few days but it did not. Instead, it gradually faded from sight! The star was next seen three months later, and subsequent observations established that it erupts every 3–4 months. Clearly it was not a normal nova and so it was that U Geminorum became the first recognised of the dwarf novae class of stars.

Typical UG-type systems consist of a semi-detached binary with a cool (spectral type K or M) dwarf or subgiant star very closely orbiting a white dwarf. The orbital periods are typically in the range of an hour to half a day. The matter streaming across from the secondary star creates an accretion disk around the white dwarf and it is this accretion disk which is the major source of light emitted by the system. X-radiation is also commonly detected from such systems, indicating sites of matter (actually the bright spot) at a particular high temperature. There is more about X-ray emitting sources in Section 10.7.

UG stars are subdivided into three main types: *UGSS* (named after SS Cygni), *UGSU* (named after SU Ursae Majoris) and *UGZ* (named after Z Camelopardalis). The light-curves of these three exemplar stars are shown in Figure 10.1. You will find detailed explanations of the distinctions between these subtypes in the 'Variability types' document on the accompanying CD-ROM. Here I just note a few defining characteristics.

Sometimes the normal outbursts of UGSU stars, similar in nature to those of the UGSS stars, are replaced by *superoutbursts* (also known as *supermaxima*) of long duration and about $0^m.5$–1^m brighter than the normal outbursts, during which there is a periodic small amplitude (circa $0^m.3$) superposed on the light-curve. These superimposed small amplitude excursions in brightness are known as *superhumps*. Superhumps have periods which are slightly longer than the established orbital period of the system.

UGZ stars are characterised by occasional sudden interruptions to their usual brightness changes. The brightness of the system becomes frozen at a level a little less than the maximum for the system. These brightness standstills can persist for the duration normally taken for several outburst–quiescent cycles. Figure 10.2 compares the long-term light-curves for SS Cygni and Z Camelopardalis. Notice the standstills in the variability of the latter star, typical for those of the UGZ subtype.

Just as the accretion disk is the major source of radiation from the system, so it is the major reason for the brightness variations. It seems that accretion disks can be unstable things. Theorists have developed a number of models, involving a number of proposed mechanisms to explain what goes on in them. As matter piles in from the secondary star, the disk is highly turbulent and mechanically unstable, with material spilling outwards and inwards. There are also factors of viscosity and the influence of the magnetic fields generated by ionised matter in motion to take into account.

You can appreciate that the astrophysics involved is highly complex and so it is of little wonder that the researches are still ongoing. However, one factor that is likely to be very important in explaining the outbursts is the fact that at a certain critical temperature, roughly 7000 K, hydrogen gas can easily switch from its neutral state to its ionised state – and it soaks up a considerable amount of energy while it is making the transition.

10.2 UG (U Geminorum) stars, aka dwarf novae

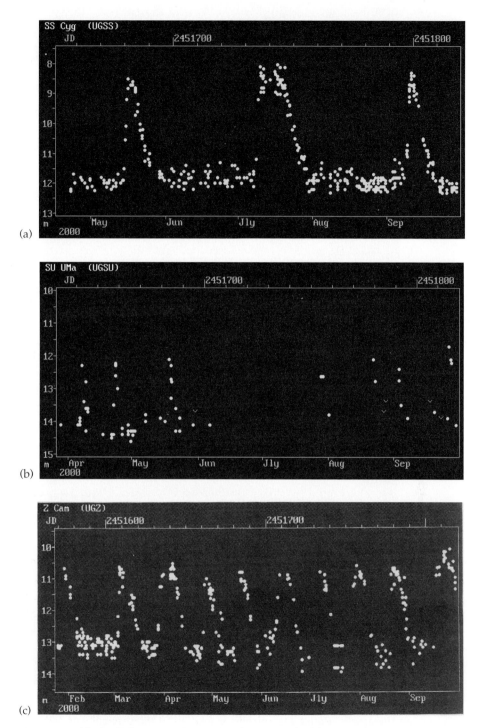

Figure 10.1 (a) The light-curve of SS Cygni (type UGSS); (b) the light-curve of SU Ursae Majoris (type UGSU); (c) the light-curve of Z Camelopardalis (type UGZ). Courtesy *BAAVSS*.

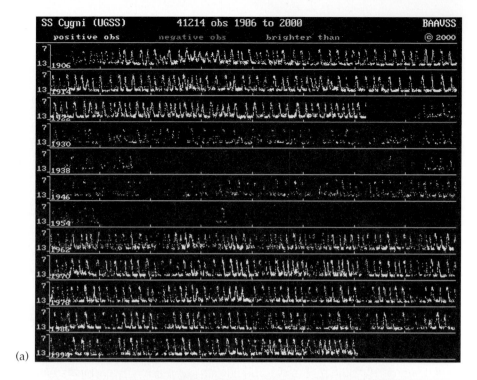

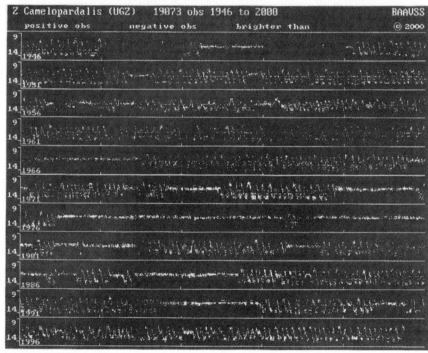

Figure 10.2 The long-term light-curves of two dwarf novae compared: (a) SS Cygni; (b) Z Camelopardalis. Notice the episodic standstills in the variability of Z Camelopardalis, typical of UGZ stars. Courtesy *BAAVSS*.

10.2 UG (U Geminorum) stars, aka dwarf novae

Also the neutral hydrogen remains uncoupled to the local magnetic fields in the accretion disk but as soon as it is ionised it becomes intimately coupled to the magnetic fields and, indeed, can influence them. The inner parts of the accretion disk are moving faster than the outer parts (a normal consequence of orbital motion) and so any magnetic fields that become 'locked in' to ionised hydrogen are first sheared and then fragmented into smaller turbulent units as a result. The viscosity in the accretion disk increases dramatically because of the coupling of the matter and magnetic fields when the hydrogen is ionised.

Another effect is that the transparent neutral hydrogen becomes opaque to the passage of radiation as it undergoes the transition to the ionised state. A high opacity means that the matter can 'hold in' heat energy and so its temperature increases to a very much higher value before its radiation exchange-rate reaches thermal equilibrium (when the power input equals the power output). Un-ionised hydrogen freely radiates its heat energy and so reaches thermal equilibrium at a much lower temperature.

We still are not certain as to the precise scenario. Different mathematical models vary in the importance they give to particular individual factors and they use them in different ways. However, all of the models do predict a situation where the matter stream can create a state of pent-up mechanical, magnetic, and thermal energy in the accretion disk. In particular, the large-scale transition of neutral hydrogen to ionised hydrogen causes an eventual fairly sudden change in the effective temperature of the disk from a thermodynamically 'cool' state (circa 3500 K) to a thermodynamically 'hot' state (circa 15 000 K). It is when the disk is in this 'hot' state that we perceive the system to be in outburst.

As the pent-up energy is released, the disk cools enough to allow the hydrogen to return to its neutral state liberating its ionisation energy in the process, thus prolonging the excess energy release from the accretion disk. As the hydrogen eventually returns to its neutral state so it de-couples from the magnetic field, allowing the magnetic eddies to disappear and the disk as a whole to return to its former mechanical configuration (the disk becomes more constricted prior to outbursts and then it expands rapidly during outburst, losing material both outwards into space and inwards towards the white dwarf).

Meanwhile the now transparent hydrogen rapidly loses its thermal energy and the temperature of the accretion disk rapidly falls back to around 3500 K. So after the bulk of the hydrogen has de-ionised, the accretion disk quickly returns to its state of uneasy quiescence. Subsequently more matter streams in from the secondary star to 'top-up' the now depleted accretion disk and then eventually to set the whole thing going once again.

Variations in the rate of matter transfer between the stars may well be a factor in many systems. Indeed, it is quite likely that the increased radiance of the accretion disk itself heats the facing surface of the secondary star enough to 'boil off' more matter and so significantly increase the rate of matter transfer. Still, most theorists now think that disk instability is most likely the major factor in explaining the behaviour of dwarf novae.

Cataclysmic and symbiotic systems

10.3 Eclipsing dwarf novae

In most cases of UG systems the light we receive is a mixture of that from the secondary star, the white dwarf, the accretion disk, and the bright spot. How are astronomers to untangle the mixture in what they perceive to be a point of light and understand the physics of what is going on? Fortunately, an accident of geometry in some systems provides the opportunity for untangling the mixture.

Owing to the fact that dwarf novae consist of two very closely orbiting stars, a fairly high percentage of them undergo eclipses. This gives astronomers a wonderful opportunity to probe the structure of these systems which are ordinarily unresolvable by normal techniques.

There is an animation on the accompanying CD-ROM which illustrates the situation. Once you have the CD started on your computer click on the 'Index' folder. Under the heading 'Movies' you will see the entry 'Simulation of a CV'. Double click on that entry and the animation will start running. You will see on your monitor a large red sphere (representing the secondary star, though not with its correct shape) moving round a yellow disk. At the middle of the yellow disk is a clearing at the centre of which is a very small white sphere. The very small white sphere represents the white dwarf primary star and the yellow disk represents the accretion disk. The matter stream is not represented in this animation but the bright spot is, by a larger (though still small) white sphere moving at the outer edge of the 'accretion disk'. You will notice that the 'bright spot' continually moves just ahead of a line joining the white dwarf to the red star. This accurately reflects the real situation. The bright spot really is situated forward of the radius vector because the matter from the secondary star begins to spiral in as it approaches the edge of the accretion disk owing to the conservation of angular momentum.

As the animation runs, the view slowly changes from that looking straight down onto the system to more and more of a side-on view of the system. Notice the angle at which the eclipses begin. You can see that as the angle continues to change more and more of the accretion disk, then the bright spot, and then the white dwarf, is at first momentarily and then totally hidden by the secondary star during eclipses.

The various eclipsing dwarf novae that we know of are presented to us at their own particular angles, providing us with a selection of situations to analyse as they undergo eclipses in their quiescent states, and in their outburst states, and while they are undergoing their transitions between the quiescent and outburst states.

It is by analysing the light-curves and the changing spectra during all of these phases that we are able to isolate the physical natures of the stars, the bright spot, and the accretion disk. If you wondered how it is that we have determined that it is the accretion disk which produces most of the light in a dwarf nova, and it is in the accretion disk that the eruption happens, this is how.

10.3 Eclipsing dwarf novae

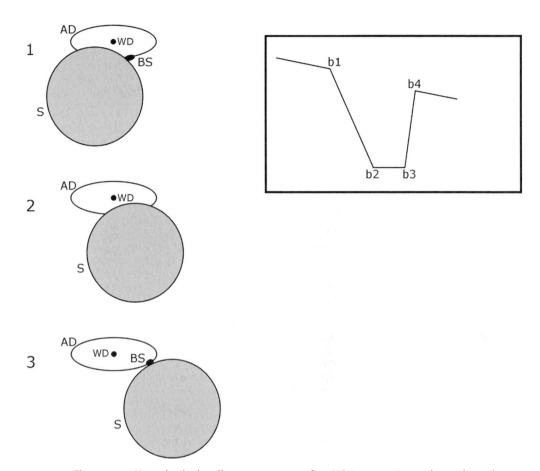

Figure 10.3 Hypothetical eclipse sequence of a UG-type system where the primary star (the white dwarf) is not eclipsed but the bright spot is, showing the light-curve this situation theoretically predicts: 1. First contact (b1) with the bright spot; 2. mid-eclipse of the bright spot; 3. last contact (b4) with the bright spot. Courtesy Nick James and W. J. Worraker.

By way of a couple of examples take a look at Figures 10.3 and 10.4. Figure 10.3 shows the sequence of events for the eclipse of a system where the angle is such that only part of the accretion disk passes behind the secondary star. Notice that the bright spot also suffers eclipse. The inset shows the effect on the light-curve for this system. Figure 10.4 shows the same for the rather more complicated situation where the angle is such as to cause the white dwarf and most of the accretion disk, as well as the bright spot, to suffer eclipse by the secondary star.

Remember, it is not just the light-curve which gives information. The way the spectrum changes gives vital clues as to what is going on. If the brightness and spectroscopic changes are monitored with sufficient time resolution, then it is

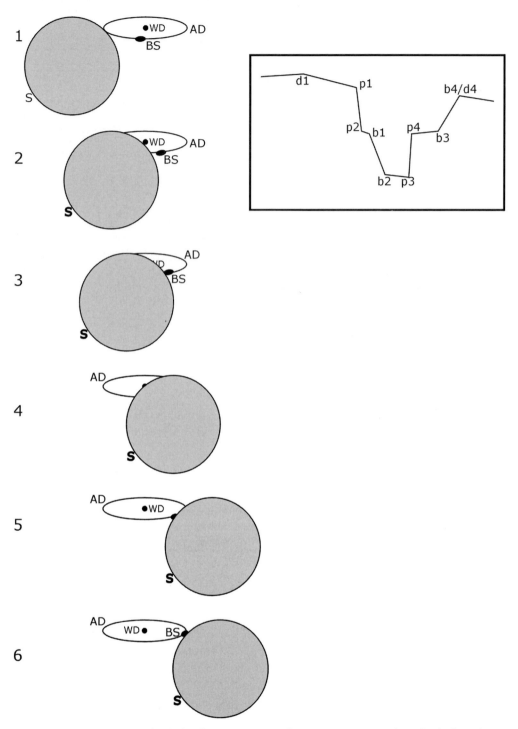

Figure 10.4 Hypothetical eclipse sequence of a UG-type system where both the primary star (the white dwarf) and the bright spot are eclipsed, showing the light-curve this situation theoretically predicts: 1. first contact (d1) with the disk; 2. first contact (p1) with the white dwarf; 3. first contact (b1) with the bright spot; 4. last contact (p4) with the white dwarf; 5. third contact (b3) with the bright spot; 6. last contact (b4/d4) with the bright spot and disk. Courtesy Nick James and W. J. Worraker.

even possible to gauge the physical sizes and locations of the various emitting regions.

High time-resolution photometry has even allowed us to deduce that the change of state of an accretion disk at the onset of eruption often initiates at its outer edge and spreads inwards as a rapidly moving wavefront. This is an outside-in or *type A* eruption. Sometimes, though, the eruption starts at the inside edge of the disk, closest to the white dwarf, and spreads outwards. This inside-out situation is termed a *type B* eruption.

One example of a star showing type B eruptions is IP Pegasi. You will find light-curves and charts for this star on the accompanying CD-ROM. You will also find an image and a graph of a photometric run on it while it underwent an eclipse under the 'Various images' heading on the CD-ROM. There is also a time-lapse movie of IP Pegasi undergoing an eclipse on the CD-ROM. You will find it under the 'Movies' heading. Just double click on it to start it going.

The following section details the dwarf novae content of the accompanying CD-ROM – but before leaving this subject, I should like to mention a very good paper in the April 2003 (Vol.113 No.2) *Journal of the British Astronomical Association*. The paper is entitled 'Eclipsing dwarf novae' and is by W. J. Worraker and Nick James. If this subject interests you, I recommend that you look out this paper. It provides a very detailed grounding, discusses several examples of eclipsing dwarf novae systems, and the authors conclude by suggesting various lines of research that the advanced practitioner can pursue.

10.4 Dwarf novae on the accompanying CD-ROM

The somewhat erratic and unpredictable behaviour of any one dwarf nova at any time and the complexities of the light variations that arise as a result of the various (and to some extent varying) conditions at each system make these especially fascinating subjects for the observer. You will never quite know what to expect when you turn your telescope towards any UG star. The monitoring of dwarf novae is probably one of the most valuable of all the types of variable star observation you can undertake.

As you might expect, the entries for this type of star outnumber those for any other on the accompanying CD-ROM. In fact, there are so many light-curves of dwarf novae on the CD-ROM that I have split them here under the three headings that you will find in the 'Lightcurves' document:

Preliminary light-curves from recent BAA/TA reports

RX And (UGZ); AR And (UGSS); DX And (UGSS); FN And (UGSS); FO And (UG); DH Aql (UGSS); SS Aur (UGSS); TT Boo (UGSU); HL Cma (UGSS); SV Cmi (UGZ); Z Cam (UGZ); AF Cam (UG); AM Cas (UGSS); DK Cas (UGSS); GX Cas (UGSU); HT Cas (UGSS + EA); KU Cas (UGSS); V452 Cas (UGSU); WW Cet (UGZ); SY Cnc (UGZ); YZ Cnc (UGSU); AK Cnc (UGSU); AT Cnc (UGZ); TT Crt (UGSS); SS Cyg (UGSS); EM Cyg (UGZ + E); V503 Cyg (UGSU);

Cataclysmic and symbiotic systems

V516 Cyg (UGSS); V630 Cyg (UGSU); V1028 Cyg (UGSU); V1060 Cyg (UGSS); V1504 Cyg (UGSU); AB Dra (UGZ); EX Dra (UG + E); U Gem (UGSS + E); IR Gem (UGSU); AH Her (UGZ); V844 Her (UGSU); RZ LMi (UGSU); X Leo (UGSS); AY Lyr (UGSU); CY Lyr (UGSS); DM Lyr (UGSU); LL Lyr (UG); V493 Lyr (UG); CN Ori (UGZ); CZ ORI (UGSS); V1159 Ori (UGSU); RU Peg (UGSS); HX Peg (UGZ); IP Peg (UGSS + E); V368 Peg (UGSU); TZ Per (UGZ); FO Per (UGZ); KT Per (UGZ); QY Per (UGSU); TY Psc (UGSU); NY Ser (UGSU); RZ Sge (UGSU); SU UMa (UGSU); SW UMa (UGSU); BC UMa (UGSU); BZ UMa (UG); CH UMa (UGSS); CY UMa (UG); DI UMa (UG); ER UMa (UGSO); SS UMi (UGSU); and VW Vul (UGZ).

Long-term 'pixel' light-curves

RX And (UGZ); UU Aql (UGSS); SS Aur (UGSS); SS Cyg (UGSS); EM Cyg (UGZ + E); AB Dra (UGZ); IR Gem (UGSU); AH Her (UGZ); X Leo (UGSS); AY Lyr (UGSU); CN Ori (UGZ); CZ Ori (UGSS); RU Peg (UGSS); TZ Per (UGZ); SU UMa (UGSU); SW UMa (UGSU); and CH UMa (UGSS).

Other detailed light-curves

RX And(1) (UGZ); RX And(2) (UGZ); DX And (UGSS); FN And (UG); Z Cam (UGZ); HT Cas (UGSS + EA); SS Cyg (UGSS); EY Cyg (UGSS); V542 Cyg (UGSS); V1113 Cyg (UG); V1251 Cyg (UGSU); DO Dra (UG); U Gem(1) (UG); U Gem(2) (UG); IR Gem (UGSU); UV Per (UGSU); RZ Sge (UGSU); S UMa (UGSU); CY UMa (UG); and HV Vir (UG).

Of these, the following have TA charts also on the CD-ROM:

RX And; FN And; HL Cma; AF Cam; HT Cas; V452 Cas; SY Cnc; YZ Cnc; AK Cnc; AT Cnc; EM Cyg; V503 Cyg; V516 Cyg; V630 Cyg; V1028 Cyg; V1504 Cyg; IR Gem; V844 Her; DM Lyr; LL Lyr; V493 Lyr; V1159 Ori; IP Peg; QY Per; TY Psc; RZ Sge; SW UMa; BC UMa; BZ UMa; CY UMa; DI UMa; ER UMa; VW Vul; DX And; FN And; EY Cyg; V542 Cyg; V1113 Cyg; V1251 Cyg; UV Per; HV Vir; and V344 Lyr.

There are also TA charts of the following additional stars on the accompanying CD-ROM (the types are given in parentheses, with a question mark in cases of uncertainty):

HP And (UG?); LL And (UG); XY Psc (UG?); WX Cet (UG?); TT Ari (UGZ); UW Per (UG?); NSV00895 (UG?); UW Tri (UG?); HW Tau (UGSS); V344 Ori (UGZ); CW Mon (UGSS); AQ CMi (UG?); AW Gem (UGSS); EG Cnc (UG); AK Cnc (UG); TU Leo (UG?); DV UMa (UG); RU LMi (UG); SS LMi (UG?); RZ leo (UG?); T Leo (UGSU); DO Dra (UG); AL Com (UG); EX Hya (UGSU); GO Com (UGSS); VZ Boo (UG); V419 Lyr (UG?); V795 Cyg (UGSS); V811 Cyg (UGSS); V1454 Cyg (UGSS); V725 Aql (UG); AW Sge (UG); V337 Cyg (UG?); V550 Cyg (UGSS); V1363 Cyg (UGZ); V1316 Cyg (UGSU); HO Del (UG?); TY Vul (UG); VY Aqr (UGSU); EF Peg (UG); EG Aqr (UG); and V630 Cas.

10.5 Other cataclysmic subtypes

Of those on the light-curves list, the following also have BAAVSS charts on the accompanying CD-ROM:

SS Aurigae; V1008 Herculis (UG?); RX Andromedae and IP Pegasi.

There are, in addition, BAAVSS charts for TT Crateris (UG) and V2176 Cygni (UGSU).

There are also the following light-curves of probable UG stars on the accompanying CD-ROM, though their pedigree is less than certain:

LD317 And; CI Aql; V1008 Her; LX Ser (EA + UG?); QW Ser; SBS1017+533 UMa; FY Vul; and EV Aqr.

Of these, there are TA charts of CI Aql and LX Ser, and a BAAVSS chart of V1008 Her.

Finally, there are the following images of UG (dwarf novae) stars on the accompanying CD-ROM:

V1060 (type UGSS); V1363 Cyg – plus a frame showing photometry of this star; V1251 (UGSU); V452 (UGSU); V503 Cyg (UGSU); V516 Cyg (UGSS) photometry; V811 Cyg (UGSS); AL Com (UGSS); EM Cyg (UGZ + E) – image and frame of photometry during an eclipse; HT Cas (UGSS + EA) – image plus a frame of photometry during an eclipse; IP Peg (UGSS + E) – image plus a frame of photometry during an eclipse; KV And (UG) – image plus frame of photometry; QY Per (UGSU); SS Aur (UGSS + E); U Gem – image plus two frames of photometry; UV Per (UGSS); V1028 Cyg (UGSS) – two images showing different states.

10.5 Polars, intermediate polars, and other cataclysmic subtypes

The GCVS 'Variability types' document on the accompanying CD-ROM lists the various types of cataclysmic variable stars as: 'N, NA, NB, NC, NL, NR, SN, SNI, SNII, UG, UGSS, UGSU, UGZ, ZAND'. SN, SNI, and SNII variables will be discussed in the next chapter, all of the others are covered in either this chapter or Chapter 9.

While I recommend that you should stick to the GCVS classification scheme wherever possible (and especially while you are still finding your way around the subject), you should be aware that other authorities have their own classifications and subdivisions. Of relevance here are a few very commonly encountered subtypes of cataclysmic variable stars. These may include *AM Her, DQ Her, UX UM a, VY Scl* stars, plus a few others. The ones I have listed here are often regarded as subtypes of nova-like variables (NL stars, see Section 9.7), though the GCVS classification for the NL class of object is really for cataclysmic variables of less well-defined character.

AM Herculis (AM Her) stars are of great interest to astronomers as the white dwarf component of these systems are very highly magnetised. This magnetic

Cataclysmic and symbiotic systems

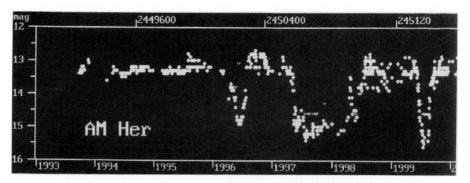

Figure 10.5 The light-curve of AM Herculis. Courtesy BAAVSS.

field dominates the behaviour of the plasma stream emanating from the secondary star. In particular, it disrupts the stream as it nears the white dwarf, preventing the formation of any really substantial accretion disk. Instead the stream splits into two and is funnelled along the magnetic field lines down to the poles of the white dwarf. Thus a cloud of accreting matter is formed over each pole of the white dwarf.

The normal behaviour that we observe is a small amplitude flickering and longer-term periods of high and low brightness. These stars tend to be rather faint and they mostly display only small amplitude brightness fluctuations, so they are not good quarry for us. However, you will find light-curves for three examples: BY Cam; AM Her (two examples), and HZ Her, plus a TA chart of BY Cam on the accompanying CD-ROM.

Actually AM Herculis is one example where eclipse effects provide us with larger than typical brightness fluctuations for this class (see Figure 10.5). This star is classified as AM/XRM + E. AM Herculis stars are best known as *polars*, because of the dominating effects of the powerful magnetic fields of their white dwarfs.

DQ Herculis stars are best known as *intermediate polars* because, as you might imagine, their white dwarfs have unusually strong magnetic fields, though rather less so than in AM Her systems. They show characteristics halfway between those of polars and 'normal' cataclysmic systems. So in these stars we see evidence for an accretion disk as well as accretion clouds over the poles of the white dwarf.

Again the dearth of sufficiently bright examples means that they are not really for us. For these and other reasons, I do not propose discussing any of the other sometimes-used subtypes of cataclysmic variable here.

10.6 ZAND (Z Andromedae) stars

ZAND stars, named after the prototype *Z Andromedae*, are a rather inhomogeneous – and consequently rather loosely defined – class of interacting stars. They are often known as symbiotic stars. If there is a typical ZAND star, it will consist

10.6 ZAND (Z Andromedae) stars

of a hot main sequence star very closely orbiting a cool giant star. The mass transfer that occurs from the giant star (the secondary) to the hot star (the primary) tends not to produce any really substantial accretion disk, though there is some evidence for such a disk in some ZAND systems. Some systems *might* have a white dwarf as the primary star, in which case one would expect a more substantial accretion disk.

Spectroscopically, these stars are notable because they show evidence of greatly differing temperature regimes coexisting in the one system. There are absorption bands from the secondary star and even cooler matter which must be below about 4000 K to form, as well as intense emission features which are evidence for temperatures above 100 000 K (at the sites of the interactions with the matter stream).

ZAND stars are also notable for evidence for material lost to the secondary star spread beyond the Roche lobes and into the whole volume of space occupied by the system (as mentioned earlier, this is known as a common envelope). This gaseous envelope is excited into emission by the radiation from the hot star. Of course the shape of this emission halo appears to change as this system rotates. The matter stream from the secondary star also creates a hot spot at the photosphere of the secondary star, or above its photosphere, or at its accretion disk if there is one, or at some or all of these sites. There might be variations in the distribution of matter between these sites. The flow of matter from the secondary star might itself be variable.

As you might imagine, this complicated situation results in highly erratic variations in the light output of the systems, sometimes amounting to several amplitudes. They are fascinating stars and there are many examples you can follow with modest equipment and by purely visual means.

You will find light-curves for the following ZAND stars on the accompanying CD-ROM (where these are of mixed type, their classifications are given in parentheses):

Z And; CH Cyg (ZAND + SR); YY Her; BX Mon (ZAND + E); CH Cyg (ZAND + SR); CI Cyg (EA/GS + ZAND); AX Per; Z And; EG And; V1413 Aql (ZAND + E); CH Cyg (ZAND + SR); CI Cyg (EA/GS + ZAND); AG Dra; YY Her; and AX Per.

Remember, the duplications in the foregoing list are where alternative light-curves are provided. There is another system, QW Sge, for which the classification is not certain at the time I write these words.

Of the above list, there is a TA chart on the accompanying CD-ROM for AX Per. There is also a TA chart for the system V1017 Sgr, which may or may not be a true ZAND star. From the same listing, the following systems are represented by BAAVSS charts on the accompanying CD-ROM:

Z Andromedae; CH Cygni; YY Herculis; BX Monocerotis; BF Cygni; AX Persei; EG Andromedae; and AG Draconis.

Cataclysmic and symbiotic systems

There are also BAAVSS charts for NQ Geminorum (SR + ZAND) and TX Canum Venaticorum on the accompanying CD-ROM.

10.7 Intense X-ray sources

You will remember from Chapter 5 that continuum radiation from a source varies in its power output, and in its wavelength of peak emission, with temperature. By the time the temperature of any body climbs to a million degrees or more it is emitting very strongly at wavelengths short enough (circa 1×10^{-10} m) for the radiation to be called X-rays.

When the particles in a matter stream from a donor star slam into an accretion disk, they often heat the bright spot to a temperature of a million degrees or more and so become a source for strong X-ray emissions. If the primary star is more compact than a white dwarf, say a neutron star or even a black hole, then clearly the potential is there for even more energetic interactions owing to its more concentrated gravitational field.

So it is that the conditions in many interacting binary systems produce a sufficiently high flux of X-rays to be defined as *intense X-ray sources*. On the 'Variability types' document on the accompanying CD-ROM you will find that the GCVS have given such X-ray sources their own classification. Of course, we are only able to observe the visible light variations that accompany the energetic phenomena (not many amateurs are going to have their own orbiting X-ray satellite with which they can do their observations!). The GCVS classification of *Optically Variable Close Binary Sources of Strong Variable X-ray Radiation* is split into the defined types: *X, XB, XF, XI, XJ, XND, XNG, XP, XPR, XPRM* and *XM*. You will find concise definitions for each of these types in the 'Variability types' document on the accompanying CD-ROM.

Involving as they do particularly energetic processes, the study of intense X-ray sources is truly fascinating. However, as far as suitable observational quarry goes, most are less than ideal for the simply equipped amateur or the beginner. They tend to be faint and/or have small brightness amplitudes, so I am not going to elaborate on them further in this present book. I can, however, offer you a few examples of light-curves and charts that may be of interest and help in getting you started if you do wish to take up their study and observation in the future.

You will find examples of light-curves for the following systems on the accompanying CD-ROM, with the specific types given in parentheses:

V635 Cas (XNG); AM Her (dually classified as AM and XRM + E); XTE J1118+480 UMa (XT); X Per (GCAS + XP); and GK Per (NA + XP, also commonly regarded to be an intermediate polar).

Of these just V635 is represented in the TA charts listing, though there are also TA charts of the following additional systems:

10.7 Intense X-ray sources

BY Cam (AM/XPRM?); V616 Mon (XND); AN UMa (E + XRM); and V1343 Aql (E + XJ).

There are images of HT Cam (type XM) and the X-ray nova XTE1118+480 UMa in the 'Various Images' document on the accompanying CD-ROM.

In the next, and final, chapter of this book we encounter astrovariables of the greatest energies and violence of all.

Chapter 11
The extra-galactic realm

In this chapter we will be considering the most extreme examples of cataclysmic variables – one-off events in which the star is destroyed. In the least powerful examples of these the progenitor blows itself to pieces, briefly outshining the light of all the stars combined in its host galaxy. In the most powerful phenomena the star begins to explode but then is sucked out of time and space in a blaze of radiation that can be visible from way across the Universe. It is fortunate indeed for us that we have no candidates for these types of events in our own celestial backyard as the radiations from them would wipe out most of the life on our planet!

First, though, we will examine in some detail a type of super-dense star of relevance to our studies.

11.1 Neutron stars

White dwarfs can exist only up to a limiting mass of 1.4 times that of our Sun. This is the *Chandresekhar limit*, named after the scientist who developed the theory of it. If the dead star is of still greater mass, then electron degeneracy pressure cannot withstand the crushing forces of the star's great weight.

Such a stellar relic would continue to collapse until electrons 'tunnel' into protons, creating neutrons. It used to be thought that the resulting object would almost entirely consist of neutrons and so was given the moniker *neutron star*. Crushing the mass of a star into a diameter of less than 20 km produces a density of the order of 1×10^{18} kg m^{-3} – a hundred million million times the density of lead!

Theorists agree that the neutron star is overlaid with a solid iron crust but as to what lies beneath, well, here opinions differ widely. Some say that the sea of neutrons extends down a couple of kilometres but then the intense pressure causes the neutrons to begin splitting into their component particles (quarks). With increasing depth there are less and less neutrons as more are split into quarks. Near the centre of the neutron star all the matter would be in the form of quarks.

That is one idea. Another is that the pressure splits the neutrons totally, beginning right under the iron crust (and separated from it by a thin layer of electrons and powerful electric fields) and extending downwards. Further, the quarks that are liberated are of the type known as 'strange', which physicists postulate will exists under conditions of extreme pressure. If that scenario is true, then perhaps we should rename these bizarre stellar remains 'quark stars', or even 'strange quark stars'.

One thing that is certain is that neutron stars contain much of the angular momentum of the original star. The precursor, though rotating slowly, had a lot of angular momentum owing to its great size. As a consequence of its contraction, the neutron star must rotate very fast. Newly formed neutron stars spin on their axes around 1000 times every second. However, an object on the surface of a neutron star would not be flung off into space thanks to a surface gravitational pull over 200 million million times stronger than that on the surface of the Earth.

The magnetic field of the original star is similarly concentrated. It has a value about a million million times stronger than the Earth's field. Theorists think that the temperature is around 10 million K throughout most of the neutron star's bulk.

The incredibly powerful magnetic field must affect the photosphere of the neutron star, though the photosphere is only a few centimetres thick because of the intense gravitational pull. Much of the radiation from the star, as well as electrified particles, must preferentially escape from the regions of the magnetic poles, where the lines of magnetic force enter the photosphere nearly perpendicular to it.

Interestingly, the magnetic and rotation axes may be inclined to each other, giving rise to a 'flashing' effect as the beam of a rotating neutron star sweeps across our line of sight. These flashes, or 'pulses' have actually been observed in many wavebands. The first radio pulses were discovered in 1967 and led to these objects being called *pulsars*. They have periods ranging from 0.006 seconds to 4.3 seconds. Theory soon linked these radio objects to neutron stars, and we are now sure that they are one and the same.

11.2 Supernovae

On 1054 July 4 Chinese astronomers witnessed the sudden appearance of a brilliant star in the constellation of Taurus. The few surviving reports of the time state that the object was pinkish in colour and that it could be seen in broad daylight for three weeks. Twenty-one months passed before the object dimmed enough to be invisible in the night sky.

Figure 11.1 The Crab Nebula imaged by the author using the STV camera and 0.5 m reflector of the Breckland Astronomical Society.

In 1731 the English amateur astronomer John Bevis found a dim nebula in the constellation of Taurus. The French comet hunter Charles Messier rediscovered the object in 1758. He thought he had found Halley's comet, which was predicted to make an appearance in that year. Historians say that it was this error which prompted Messier to draw up his list of objects which might be mistaken for a comet. Certainly the nebula in Taurus is number 1 in Messier's catalogue. It is now known that the bright star reported by the Chinese was a *supernova* and the nebula M1 is the wreck of that old star. It is a *supernova remnant*.

Lord Rosse's 1.8 m aperture reflector ('The Leviathan of Parsonstown', now reconstructed in its old home in the grounds of Birr Castle in Ireland) revealed the nebula's complex filamentary structure in 1845. This structure led to the object being nicknamed the Crab Nebula, a name which we still use today.

Figure 11.1 shows a modern CCD image of the Crab Nebula taken with a much smaller telescope. Its distance is reckoned at 1800 parsecs, so the actual explosion happened about 6000 years before the Chinese observed it. The nebula is expanding at the rate of about 1500 km s^{-1} and is the source of powerful emissions of radiation in radio and X-ray wavelengths. The light from the nebula is strongly polarised, indicating that it is created by electrons spiralling along magnetic field lines and losing energy by a process known as *synchrotron emission*.

A pulsar (neutron star) was found flashing in the heart of the Crab Nebula in 1968. It has a period of 0.03 seconds, though this is slowing down at a rate of 4×10^{-13} second per second. It is the liberation of the neutron star's rotational energy that keeps the nebula emitting in visible and other wavelengths. Without this energy source the nebula would have thinned out enough to become invisible long ago.

The 1054 supernova was not the first to be observed, nor was it the last. On average one supernova is reckoned to occur in our Galaxy every 50 years, though we rarely see them because of obscuration by interstellar matter. Neither is the Crab Nebula the only supernova remnant to be seen in the sky. We see these stellar relics all over the sky. It is from the runaway nuclear reactions in

11.2 Supernovae

the supernova blast, seeded into the cosmos via the supernova remnant, that the full range of chemical elements originates. We would not exist if it were not for past supernovae.

Supernovae are of two distinct types. *Type I* supernovae are thought to be the result of mass transfer processes between a binary pair of stars, where one of the stars is a white dwarf. As is the case for novae, matter is transferred from a swollen star onto a white dwarf, via an accretion disk. If the white dwarf collects enough matter for its mass to exceed 1.4 times the mass of the Sun it cannot remain as a white dwarf. Starting at its core, it collapses to form a neutron star. This collapse is sudden and extremely rapid, leading to colossal outpourings of heat energy and neutrinos (the bulk of the energy is released via neutrinos) and runaway nuclear reactions in the outer zones of the star.

The core collapse takes just seconds and creates an outwardly expanding immense shock wave. The star blows itself apart, reaching an absolute magnitude of -19^m in a day or two after the initial collapse. The surviving core of the star is compressed by the kickback to leave the final neutron star end-product of the stellar collapse. After reaching its peak brilliance, the brightness of the supernova rapidly falls to around absolute magnitude -16^m over the next month and then continues to fall at a fairly steady rate of about 1 magnitude every 2 months.

The spectra of the minority of Type I supernovae show a significant deficiency of hydrogen compared to the majority of them. These are termed *Type Ib* supernovae, the majority of the Type I supernovae being denoted *Type Ia*. There are other differences between Type Ia and Type Ib supernovae. In particular, the Ib versions seem to involve progenitor systems of greater total mass than is the case for the Ia subtype. Some authorities also recognise other types, so you may come across Type Ic, Type III, Type IV and other subtypes of supernovae in the literature – but I do not propose elaborating on them here.

Since the Type Ia supernovae all result from the detonation of a white dwarf of 1.4 solar masses, they all reach very similar maximum brightnesses during eruption. They are also readily identified thanks to characteristics of their spectra and their light-curves. That makes them good 'standard candles' for determining the distances of their host galaxies at extreme distances.

When astronomers compared the distances of Type Ia supernovae inferred from their apparent brightnesses with the distances as indicated by their spectral redshifts, it was apparent that there was a discrepancy which increased with distance. The supernovae that were most distant (the distances as implied by redshifts calibrated on nearer objects) appeared to be systematically under-luminous. This implies that the redshift calibration for local space (and so recent look-back time) is not valid at cosmological distances (and consequently for the Universe of billions of years ago). So, it is thanks to Type Ia supernovae that we now know that we live in a Universe of accelerating expansion – and all that this revelation entails.

The extra-galactic realm

Figure 11.2 (a) and (b) Supernova SN1993J UMa in the galaxy M81 photographed by Martin Mobberley using his 0.49 m reflector; (a) the supernova (arrowed) at maximum light (about 11^m) on 1993 April 11; (b) the view on 1994 January 17. Notice that the supernova is now much dimmer, despite this second exposure showing much fainter details in the host galaxy. (c) The light-curve of this supernova as constructed from the observations of BAA members. Courtesy BAAVSS.

11.2 Supernovae

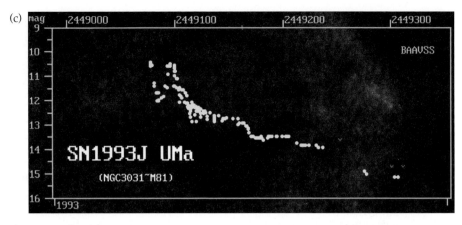

Figure 11.2 (cont.)

Type II supernovae are the final death throes of massive stars that have come to the end of their supplies of fusible material, as already described in Section 6.8. These supernovae show some diversity in their behaviour, though their common characteristics can be described. Typical Type II events are rather less bright than Type I. They most often reach absolute magnitude values of around -17^m a day or so after the core collapses and they often show a rapid fall of around a magnitude or two over the next month, remain steady in brightness for a couple of months, then rapidly fall in brightness over the following year or so. Astronomers can distinguish between Type I and Type II supernovae by the characteristics of their spectra, as well as by their light-curves.

The last supernova to be observed in our own Galaxy was seen in the constellation of Ophiuchus in 1604. However, astronomers have witnessed many of these spectacular stellar deaths in other galaxies. See Figures 11.2 and 11.3 for a couple of examples. The closest supernova of recent years happened in a satellite system to our own – the Large Magellanic Cloud. It was discovered on 1987 February 24 by a number of independent observers.

This supernova set astronomers something of a puzzle. The way its brightness changed with time did not conform to any of the previously documented cases. After discovery it gradually rose in brightness from magnitude 5^m to 3^m over a period of about 3 months. After that its brightness dwindled pretty much in the way expected for a Type II supernova. The Large Magellanic Cloud is 50 000 parsecs from us, so the absolute magnitude of the object was calculated to be $-15^m.5$. This was rather faint for a supernova, even one of Type II.

Astronomers were initially very surprised when investigation of pre-discovery images of the region showed that the progenitor star was a blue supergiant, and not the expected swollen red supergiant. However, this revelation does explain the supernova's behaviour.

The extra-galactic realm

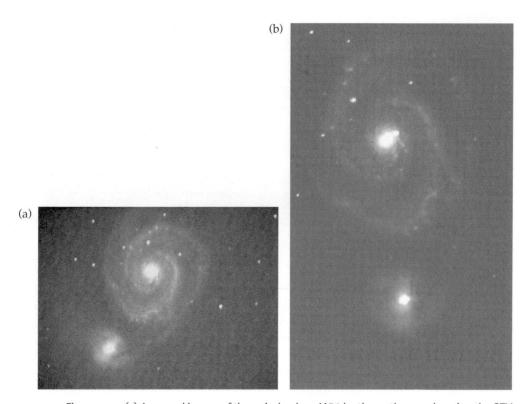

Figure 11.3 (a) A normal image of the spiral galaxy M51 by the author made using the STV camera and 0.5 m reflector of the Breckland Astronomical Society. (b) Supernova SN1994 I imaged by Martin Mobberley on 1994 April 11 using a Starlight Xpress FSX CCD camera on his 0.49 m reflector. The supernova can be seen very close to the nucleus of the galaxy and Martin has processed his image with a logarithmic stretch in order to clearly show it. Courtesy Martin Mobberley.

Turn back to Chapter 5 and the H–R diagram shown in Figure 5.5. Then find the location of the supergiants, extending from blue at the upper-left to red at the upper-right. Compare this with Figure 6.10 which shows stellar radii, and you will see that blue supergiants are physically very much more compact than red supergiants. The progenitor of this supernova was probably about 20 times the diameter of our Sun, as opposed to the more usual several hundred solar diameters.

The relatively small size of the progenitor resulted in much of the initial heat energy released by the explosion being used to expand the star until its outer layers were thin enough to allow the pent-up energy to radiate more freely. This process took the first 2 or 3 months after the initial core collapse and this is why the supernova was dimmer than usual at the beginning and why it gradually brightened over that time.

Several neutrino detectors operating at various locations on the surface of the Earth recorded bursts of activity at exactly the same time – 7^h 35^m UT on the day of discovery. Theorists think that this marked the time of the core collapse that caused the outburst (bear in mind that both the light and the neutrinos took 160 000 years to cross the space separating the Large Magellanic Cloud and the Earth, so the actual event was 160 000 years earlier). Astronomers were able to learn much about supernovae in general just because *SN 1987A*, as it is called, was so atypical.

11.3 Supernovae on the accompanying CD-ROM

The following light-curves of supernovae are to be found on the accompanying CD-ROM:

SN1999c1 Com; SN1999em Eri; SN1989B Leo; SN1998bu Leo; SN1993J UMa; SN1988S UMa; SN1998aq UMa; SN1999by UMa; SN1991T Vir; and SN1994D Vir.

Though you may think that it is akin to continuing to stare at the stage after the play is over, there is some value in monitoring a supernova's continuing decline into obscurity. To that end there are eighty TA charts of these objects on the accompanying CD-ROM. They are all faint (typically 15^m or fainter) and suitable only for the observer equipped for CCD photometry. I will not list them here but you can find them for yourself very easily. Just look down the listing for every entry beginning 'SN'!

11.4 Supernova hunting

My comments follow those for hunting novae (see Section 9.8). In particular, I must emphasise that the few successful supernova hunters spend *all* of their telescope time supernova hunting. Today's most successful hunters also spend a small fortune on their equipment. Those discoverers submitting observations to the 'Nova/Supernova Patrol' run by the TA in association with the BAA are: Mirko Villi, Giancarlo Cortini, Stefano Pesci, P. Mazza, Ron Arbour, Mark Armstrong, Tom Boles, Stephen Laurie, and Steven Foulkes.

The first discovery of this group was made visually (SN1991T in the Virgo galaxy NGC 4527, by Giancarlo Cortini and Mirko Villi) but in the USA the Reverend Robert Evans had already been systematically hunting for and discovering supernovae visually. In his most productive period (between 1981 and 1988) he found fourteen supernovae and co-discovered three others, mostly using his 0.4 m $f/4.5$ Newtonian reflector. He continued making discoveries after that time. His procedure was simple – look at as many galaxies as possible every clear night. He had memorised the appearance of hundreds of galaxies and so could instantly spot any star-like interlopers.

Now, though, the field is dominated by those using automated telescopes. The most successful of the TA astronomers at the time of writing is Tom Boles with nearly 50 discoveries to his name (though Mark Armstrong is not far behind). He

The extra-galactic realm

Figure 11.4 Tom Boles pictured with part of the array of equipment he uses for supernova hunting. Courtesy Tom Boles.

currently has a network of three 0.35 m Celestron Schmidt–Cassegrain telescopes each on a Paramount ME robotic mount with an Apogee AP7 CCD camera fitted to each telescope. Two of the telescopes are stationed in a roll-off roof observatory (see Figure 11.4) and one in a fully automated dome. Everything is controlled from a local-area-network of computers. Sitting at his master computer, Tom can survey over 200 galaxies per hour.

The elaborate and automated program he has developed allows him to bring up the images on the master monitor, interleaving them (blinking, as described in Section 9.8) with stock images. He sits looking at the image just taken and the stock image, one being 'blinked' against the other. The automated program can be interrupted to allow one of the telescopes to get further images of any galaxy in which a supernova is suspected (various transient flaws, including random pixel firings caused by cosmic ray strikes, cause a great many false alarms). Tom will only pass on any supernova find he is very confident of being genuine.

Can you compete with Tom's supernova-discovering factory? I could not, even if I was interested in devoting my entire time to this project. I would not be capable of staring at a screen searching blinking images for hours on end, let alone doing this into the small hours of the morning. An hour of that would leave me tired, with aching eyes and with a queasy headache. There is also the question of the finances required for a set-up like Tom's.

I will say no more about supernova hunting in this book. This is a task I am happiest leaving to others. Fortunately, if you are a member of a group such as TA you can at least get early news of all the discoveries that are made. Then you can carry out valuable observations of the changing brightnesses of the supernovae and submit your observations in the usual way.

In the 'Index' document on the accompanying CD-ROM you will find under the main heading 'Charts' the entry 'TA galaxy catalogue'. This contains the images of several hundred galaxies, often both in raw and processed form. You might find these useful to have to hand to help you identify any newly discovered supernovae.

11.5 Black holes

A dead star more massive than 1.4 times the mass of our Sun cannot be a super-dense white dwarf. It must be a hyperdense neutron star. According to our present understanding there is one state even beyond this. A stellar corpse of greater than 3.2 solar masses cannot exist as a neutron star. It will be crushed under its own weight until it is no more than a minute point of matter of virtually infinite density. Furthermore, its gravitational field will then be so strong that it will close off a region of space around itself from which nothing, not even light, can escape. This region is a *black hole*.

All the mass of the original object is concentrated at the centre of the black hole in what is called a *singularity*. At a certain distance from the singularity the gravitational field strength will be such that the escape velocity equals that of light, 3×10^8 m s^{-1}. This is the *event horizon* and its radius from the singularity is known as the *Schwarzschild radius*. Newtonian mechanics can be used to calculate the Schwarzschild radius, R_s, for an object of mass M kilograms:

$$R_s = \frac{2GM}{c^2}$$

where G is the universal constant of gravitation (6.67×10^{-11} N m^2 kg^{-2}) and c is the speed of light. Thus a 4 solar mass black hole (the mass of the Sun = 2×10^{30} kg) would have a Schwarzschild radius of 11.9 km. The more massive a black hole is, the greater is this radius. Any object falling into a black hole would be lost forever. Once it crossed the event horizon it would be completely cut off from the rest of our Universe.

The first candidate to convince astronomers of the reality of these bizarre objects involved one of the most powerful X-ray sources in the sky, Cygnus X-1, which appears to be co-orbiting a 20 solar mass blue giant star. This star shows perturbations in its proper motion which can only be caused by a companion of about 10 solar masses. However, the companion is totally invisible. If it were a normal star it would be very easily seen. It is too massive to be a neutron star and so it must be a black hole.

The intense X-ray flux is created by material from the 20 solar mass star falling onto an accretion disk before being swallowed up by the black hole itself. Under the powerful gravitational field of the black hole much energy would be liberated by the in-falling material, and this could explain the intense emission.

Cygnus X-1 was the first of many objects now associated with black holes. At the time of writing there are about a dozen stellar-sized black-hole candidates identified by the X-ray emissions from their accretion disks. Moreover, a whole panoply of once-baffling phenomena in the Universe are explainable once we admit the reality of black holes.

11.6 Hypernovae

In the 1960s the Americans sent a fleet of special military surveillance satellites into orbit around the Earth, each equipped to detect the bursts of γ-ray radiation that would arise from nuclear bomb tests. On 1967 July 2 one of the satellites did detect a burst of γ-rays, though of a peculiar signature. In time these satellites detected a dozen more of the mysterious bursts of γ-rays, each one from discreet sources spread randomly over the entire sky but it was only in 1973 that the information was cleared for release by the military and astronomers were told of the discovery.

γ-ray photons are the most energetic in the electromagnetic spectrum and it takes a physical process involving extremely high-energy transitions to produce them. Further, it takes a truly colossal amount of energy to produce a source of γ-rays that can be detected at astronomical distances.

Astronomers were intrigued and began their investigations – the hunt was on for what they then named γ-*ray bursters* (but we now just call them γ-*ray bursts*, abbreviated to *GRB*). Unfortunately they are very short-lived phenomena – the burst lasting from a fraction of a second to at most a few minutes – and they vanished as quickly as they appeared with no after effects that were indisputably detectable with the technology of the time. Just pinning down the site of a γ-ray burst proved to be difficult, let alone finding out what is going on to produce them.

A variety of rocket and balloon probes were sent aloft. The most successful and productive of these was the *Compton Gamma Ray Observatory* which orbited the Earth from 1991 to 1999. It detected γ-ray bursts at the rate of about one a day during that period. Nonetheless, when astronomers turned even their most powerful telescopes on the positions the bursts had seemed to come from they found nothing. Many thought that something might be seen if only the astronomers could set their telescopes on the burst sites very soon after the burst was detected.

11.6 Hypernovae

As the number of bursts detected by *Compton GRO* topped the hundred, it was confirmed that they were indeed positioned at seemingly random locations across the sky. That fact allowed for two alternative interpretations: either the bursts came from far beyond the limits of our galaxy, or that they were positioned very close to us, probably no more than a few parsecs away. If they were further away than a few parsecs but still within our galaxy, then the sources would surely be arranged in the sky in a pattern reminiscent of the shape of our galaxy. We would expect most of them to be concentrated along the Milky Way.

Most astronomers decided that the sources must be very close to us because the alternative meant that the γ-ray bursts would have to be incredibly powerful. In fact, if they were at cosmological distances the energy released in each burst would have to be of the order of 10^{47} joules, and maybe more. This is the energy that would be released if a body at least as massive as our Sun could be converted entirely into energy – and for that to happen in just a few seconds! If they were near the edge of our observable Universe, even this would be insufficient – and that would put these objects into the realms of violating Einstein's sacred $E = mc^2$.

Lone among scientists willing to contemplate γ-ray bursts coming from cosmological distances was Dr Bohdan Paczynski of Princeton University. However, eventually the evidence was to accrue that was to prove him correct.

The statistical distribution of bursts was compelling but what astronomers really needed was to obtain a spectrum of an afterglow so that they could determine its distance with certainty. Yet, no afterglows had been incontrovertibly seen at all.

The Italian *BeppoSAX* satellite was primarily an X-ray observatory but it was equipped with detectors which could double as position-locating γ-ray detectors. The breakthrough came when a Dutch team of astronomers using the 4.2 m William Herschel Telescope on La Palma located a dim spot of light – the faint and rapidly fading afterglow – from the position relayed from the satellite team quickly after *BeppoSAX* had detected its third γ-ray burst. This was the first genuine afterglow to be detected. Subsequently other teams recorded other afterglows, and eventually the first of a number of spectra was obtained.

From these spectra came the desired redshift measurements. The proof was conclusive. γ-ray bursts came from sources that were billions of light years away, far across the reaches of the Universe. So, the γ-ray bursts are events of colossal power and the question remains: what causes them?

We very recently have gained some clues thanks to the efforts of several teams of investigators working with satellites such as *BeppoSAX* and *Chandra* (another X-ray satellite), and with the world's most sophisticated and powerful telescopes, and utilising the most up-to-date technology. For instance we now know that most (perhaps all) γ-ray bursts occur in regions of dense molecular clouds – the birth places and nurseries of new stars – within host galaxies right out to the observable limits of the Universe.

Another clue is that a supernova that had unusually strong radio emissions and a very peculiar spectrum was observed in April 1998 a week after a γ-ray burst was detected at the same location. More recently, astronomers have found some lingering traces of supernovae light mixed in with some other γ-ray burst afterglows. Further, burst spectra analysed from the *BeppoSAX* and *Chandra* indicate the presence of iron in large quantities – forging a further link with supernovae.

A star with a mass of around ten times that of our Sun will end its life with a supernova explosion. It will blow most of its material away into space, the residue being transformed into a neutron star, as described earlier in this chapter. However, a star of still greater mass, say about 30 solar masses, would have a lifetime of 5–10 million years and would come to a different end. There would be a supernova blast but this time the kickback would concentrate more mass into the core region. If the matter concentrated was more than 3.2 solar masses then the star's core would turn into a black hole!

The consequence of that is the rest of the star would, despite the explosive reactions in its outer envelope, collapse inwards towards the black hole. The star would have been rotating so the material would spin up to extremely high velocities as it piled down to the region around the black hole. Material would vortex and stream into the black hole at each of the rotation poles in the manner of water swirling down a plug-hole. The black hole would increase in size and become ever more hungry for the remaining material.

In a matter of moments the star would become a rapidly contracting doughnut of dense, extremely hot, matter, rapidly being consumed by the 'open mouths' at each rotation pole of the black hole. Vast amounts of energy would be liberated because of the fast spinning doughnut of matter and the intense gravitational field close to the black hole. This energy would be manifest as intense electromagnetic radiation and electrified particles sprayed outwards in a thin beam, or *jet*, from each pole. The jets would slam into interstellar matter in the molecular cloud causing shock waves, heating of the matter to colossal temperatures, and intense radiation bursts.

In a few moments after the onset of the collapse it is all over – the star has gone and only a black hole remains. Meanwhile if either beam happens to be aligned in our direction we detect a γ-ray burst.

The fact that the energy from the collapsing star is concentrated into two thin beams, rather than spread in all directions as one would expect in a normal explosion, obviates having to violate Einstein's $E = mc^2$. If we see the γ-ray burst at all, then it is only because one of the beams is aligned in our direction. In that case we are seeing nearly half the total power output of the imploding star, not the minute amount we would see if the energy was radiated in all directions. So, even the γ-ray bursts at the limit of the observable Universe do not violate Einstein's equation.

Most astronomers think this is how the γ-ray bursts are formed. Astronomers have taken to calling the death throes of supermassive stars *hypernovae*. They are truly the most powerful phenomena we have so far encountered in the Universe, not counting the creation of the Universe itself.

However, we still have many details to sort out with our theoretical models. Recently, we have come to recognise that the above scenario will work quite well for γ-ray bursts of a second and longer but a minority of bursts have been detected which are much shorter in duration – a tenth of a second or even less. They appear to have a different energy spectrum from the longer bursts. So far no afterglows have been detected from the short-duration bursts. It seems likely that there are two (some speculate three) causes for γ-ray bursts: with hypernovae being the source of those whose duration is a second or longer and an as-yet unknown cause (though theories abound) for the ones of shorter duration.

Thanks to the technology now available to amateur astronomers, they have recently been able to join the professionals in even this extreme field of study. Take a look at the 'Index' document on the accompanying CD-ROM under the heading 'Miscellaneous' and you will see a document entitled 'The first UK Gamma Ray Burst detection'. There you will find a detailed account of how TA members Nick James, Mark Armstrong, Martin Mobberley, and Tom Boles were alerted by a professional detection of a GRB and were able to take valuable visual images of the afterglow of the event. The earliest image – and the first by a UK amateur – was obtained by Nick James. The account includes five images obtained on the detection day (2002 October 4) and there is one image taken two days after. There is also included Nick James' photometry of the GRB, which spanned about four hours on the day of detection. This and the images are accessed in the usual way by clicking on the links.

11.7 Quasars and active galaxies

By 1960 the infant field of radio astronomy was beginning to blossom. Radio astronomers had found and catalogued several hundred radio sources in the heavens and optical astronomers were busy trying to find their optical counterparts.

Two optical astronomers, Allan Sandage and Thomas Matthews, found a sixteenth-magnitude, bluish, star-like object at the position of the forty-eighth radio source in the Third Cambridge catalogue. 3C 48 proved to have a peculiar spectrum. It showed several broad emission lines that the astronomers could not identify with any known elements and its ultraviolet intensity was very much greater than any normal star of the same brightness. Further observations provided another surprise. It was found to fluctuate in brightness. Its brilliance changed by more than a third over the course of a year. Astronomers first thought that 3C 48 was a peculiar, and unique, radio-emitting star situated within the confines of our galaxy.

However, in 1963 another strong radio source, 3C 273, was identified with a thirteenth-magnitude bluish 'star', showing the same sort of peculiar

The extra-galactic realm

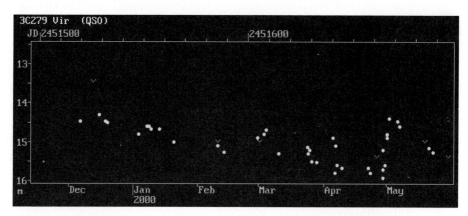

Figure 11.5 The light-curve of the quasar 3C 279, in Virgo. Courtesy BAAVSS.

spectrum. Maarten Schmidt, working at Mount Palomar, realised that the peculiar spectrum of 3C 273 could be explained as the characteristic lines of hydrogen redshifted by 15.8 per cent. In other words all the wavelengths of the spectral lines were increased by this amount. We usually refer to this redshift value as 0.158.

He took a look at the spectrum of 3C 48 and realised that its spectrum was redshifted by an even greater amount: 0.367. Such large redshift values correspond to recessional velocities that are a significant fraction of the speed of light. Clearly the objects could no longer be regarded as residing in the Galaxy. If they obey Hubble's law, then they must be immensely remote.

The distances of 3C 273 and 3C 48, turn out to be over 600 megaparsecs (2000 million light years) and over 1400 megaparsecs (nearly 5000 million light years) respectively. For these objects to appear as bright as they do despite their immense distances, they must be incredibly luminous. Over the years astronomers have discovered thousands of these 'quasi-stellar radio sources', or *quasars*. Most of them fluctuate in intensity in all observed wavelengths. There is a light-curve of 3C 273, along with others, on the accompanying CD-ROM, and Figure 11.5 shows the light-curve of the quasar 3C 279.

Investigations by Sandage showed that only a small fraction of the known quasars are strong radio emitters. However, all have very large redshifts and they are all exceedingly energetic objects. A typical quasar has a power output 100 times as great as a large normal galaxy!

One striking point about the quasars is that they are all very remote objects. This means that we are seeing them as they were thousands of millions of years ago. Indeed, quasars seem to be a phenomenon particular to the young Universe.

The fact that many of the quasars show rapid changes in optical and radio brightness indicates that the regions within the objects that produce the power output are very small; certainly less than a light year in diameter. What mechanisms could release such excessive amounts of energy in so small a volume?

11.7 Quasars and active galaxies

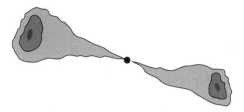

Figure 11.6 A representation of a two-lobed extended radio source. The darkest areas represent regions of the most intense emission. The lightest areas represent the regions of weakest emission.

Astronomers were puzzled. A clue came with the discovery of many galaxies that show a 'watered-down' version of a quasar's behaviour. These objects are called *active galaxies*.

As technology increased during the 1960s and beyond, so radio sources could be studied with greater sensitivity and resolution. Many were found to have a common radio structure. As well as a central emitting region whose position coincided with a galaxy, two massive 'lobes' of radio emission were often revealed to extend in opposite directions from the central object. Figure 11.6 shows the sort of radio map obtained from one of these objects. These lobes are the largest single structures in the observed Universe. For instance, the radio lobes of the source 3C 236 extend across 5 megaparsecs (over 15 million light years).

These lobes appear to be formed by millions of solar masses of plasma ejected from the galaxy at speeds that are an appreciable fraction of the speed of light. The fast-moving plasma generates a magnetic field, and the charged particles within it spiral along the magnetic field lines releasing large amounts of energy as synchrotron emission. These *extended radio sources* tend to be unvarying in their output, but the *compact radio sources* fluctuate irregularly over relatively short periods of time. They are the objects at the centre of the extended radio sources, but many compact sources exist without any apparent lobes.

Some radio galaxies exhibit jets of material extending from their nuclei. These jets are seen in visible light as well as by their radio emissions. The brightest galaxy in the Virgo cluster, M87, shows a jet of this type (see Figure 11.7). The galaxy was identified as a strong source of radio emissions in 1949, and it was given the designation Virgo A. These jets are very much smaller than the radio lobes possessed by many radio galaxies, having lengths in the kiloparsec range.

In 1943 Carl Seyfert had optically identified a number of galaxies with very bright nuclei. The spectrum of a normal galactic nucleus shows a continuum of black-body type crossed by dark absorption lines, caused by the amassed light of large numbers of stars. However, the nuclei of these *Seyfert galaxies* displayed curious spectra dominated by broad emission lines superimposed on

Figure 11.7 The jet emanating from the active nucleus of the galaxy M87. Courtesy NASA and the Hubble Heritage Team (STScI/AURA).

a continuum produced by synchrotron emission. In the following years radio astronomers found that Seyfert galaxies were powerful radio sources.

In 1968 Maarten Schmidt identified one particular compact radio source with the distant elliptical galaxy BL Lacertae. It seemed to fluctuate even more wildly than Seyfert galaxies (see Figure 11.8) and its optical spectrum proved to be very odd. Only a continuum emission could be seen, with no absorption or emission features visible at all! Other examples of these *BL Lac* objects were found and we now know of hundreds. BL Lac and Seyfert galaxies are all characterised by very bright nuclei. Many of these objects seem to be as powerful as the weakest quasars.

11.8 Cosmic chameleons

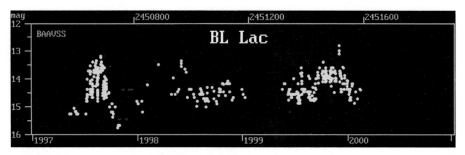

Figure 11.8 The light-curve of BL Lacertae, the 'prototype' BL Lac object. Courtesy BAAVSS.

A link between these apparently diverse objects was forged when it was discovered that many quasars are surrounded by elliptical masses of stars. In other words, quasars are the most extreme examples of galaxies with active nuclei. In the case of the quasars, the nucleus outshines the rest of the galaxy. Of course, we still had to figure out the mechanisms that produce the colossal power outputs of these objects.

11.8 Cosmic chameleons

The normal nuclear processes occurring in stars are rather inefficient in that only a tiny fraction of the mass of a star is converted into energy. A large normal galaxy would shine because of the conversion of something like 0.005 solar masses per year into energy (the total power output being of the order of 1×10^{39} watts). In order to explain the output of a typical Seyfert galaxy one would have to envisage the conversion of 0.05 solar masses per year. Quasars would require at least ten times this mass each year. Moreover, the energy-producing regions of these objects are less than a light year across.

Researchers tried to explain the power sources of these objects in a number of different ways, none of which was really satisfactory. It was only in the 1990s that most astronomers agreed on the probable mechanism. The activity in each active galaxy is caused by the presence of a supermassive black hole in the galaxy's nucleus. A black hole of a million solar masses would certainly gobble up stars and interstellar matter at a rate sufficient to release the amounts of energy observed. In the process the black hole would increase its radius, drawing in yet more matter.

Dynamical studies of the motions of stars and interstellar matter near galactic nuclei show that very large concentrations of matter *are* indeed present. Further than that, in most cases black holes are the only reasonable candidates for this mass.

Certain questions still needed answering. For instance, why are only 10 per cent of the visible quasars strong emitters of radio radiation? What produces the bland spectra shown by BL Lac objects? Why is it that not all galaxies show quasar-like behaviour? Now we have answers to those questions, too. Of

The extra-galactic realm

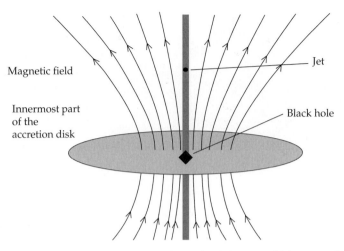

Figure 11.9 A representation of the disk of plasma orbiting very close to a supermassive black hole. The magnetic field is most concentrated near the black hole, above and below it, and in the vicinity the field lines are nearly perpendicular to the plane of the disk. It is at these locations that jets are formed, as described in the text. Beyond this plasma disk (and not shown on this diagram) is a large and much thicker 'doughnut' of cooler gas.

relevance to the last question, we now know that the centre of many apparently normal galaxies, including our own, does emit significant amounts of radio and other electromagnetic radiations.

A general theory has emerged which explains the diverse phenomena we have discussed simply as different aspects of the same thing. The key to it all is the presence of a supermassive black hole as the central power house in active galaxies.

A supermassive black hole in the centre of a galaxy will have a doughnut-shaped accretion disk of in-falling material in orbit around it. The sharp eye of the Hubble Space Telescope has actually seen a few of these galactic doughnuts, so this is no longer just supposition.

The material close to the event horizon of the black hole is at a temperature of perhaps hundreds of thousands of degrees. That in the outer, thicker, part of the doughnut is much cooler – perhaps having a temperature of a few hundred degrees. The main part of the doughnut is a couple of light years across and is very opaque.

The superhot material orbiting in a thin sheet close to the black hole is in the form of a plasma. The circulating plasma generates a powerful magnetic field (electric charges in motion naturally give rise to a magnetic field) in the manner shown in Figure 11.9. As the diagram illustrates, in the directions perpendicular to the plane of the plasma disk the magnetic field lines extend in straight lines out into space. These form 'fast-lane highways' for escaping electrified particles from the high-energy regions around the black hole. We now realise that the once mysterious jets that emanate from active galaxies are a natural consequence arising from an accretion disk around a supermassive black hole.

Of course, the material streaming (and spiralling) along the jets would plough into the surrounding interstellar medium and one might expect radio signatures of the interactions that would result. We see these, too: they are the radio lobes. Observations over the last few years have shown that the optical jets and the inner parts of the radio lobes are aligned, proving their common origin. The outermost parts of the radio lobes are often distorted and displaced, but this is to be expected from relative motions in the intergalactic medium.

So, if we see a galaxy with a highly active nucleus more or less edge-on to us then it will appear like a typically radio-noisy quasar – less active and we might call it a Seyfert galaxy.

The jets emitted by an active galaxy's nucleus would be expected to produce very bland spectra, in fact, exactly the sort that are typical of a BL Lac object. It seems likely that BL Lac objects are the same sort of active galaxy as just described. They only appear different because of their different orientations. In the cases where we look straight down one of the active galaxy's jets we call the result a BL Lac object.

If we see the active galaxy at an intermediate angle then we will observe properties between the two extreme cases. Examples of these intermediates abound. In effect, active galaxies are cosmic chameleons!

Theory also predicts that when the supermassive black hole becomes large enough, the energy streaming from it pushes the accretion disk away from the hole's event horizon. In effect, when the hole becomes too bloated it pushes its food source out of its own reach! So, at some stage the active galaxy turns into a quiescent one. Our own galaxy was undoubtedly an active galaxy in its early life. Now the 3 million solar mass black hole residing at its centre lies fitfully dormant, with only very small amounts of gas and dust occasionally stirring the gentlest electromagnetic and particle emissions from it.

In one fell swoop this theory predicts why quasars and other active galaxies are all distant from us (because they are associated mainly with an epoch long ago) and why our galaxy and most galaxies close by are quiescent, despite the recent discovery that most of them harbour supermassive black holes in their centres.

11.9 Quasars and active galaxies on the accompanying CD-ROM

A quasar, an active galaxy, a Seyfert galaxy, a BL Lacertae object – they are all manifestations of an *Active Galactic Nucleus*, or *AGN*. A few examples are of sufficient apparent visual magnitude to be seen in large amateur telescopes but the CCD revolution in amateur astronomy has opened the door to amateur imaging and photometry of them. Even so, they are really subjects for the experienced and well-equipped practitioner. Still, as observational quarry they must rank as the ultimate astrovariables!

The extra-galactic realm

The accompanying CD-ROM contains the light-curves of the following AGNs:

3C66A And (BL LAC); S5 0716+71 Cam (BL LAC); OJ+287 Cnc (BL LAC); W Com (BL LAC); BL Lac (BL LAC); NGC7469 Peg; NGC1275 Per; BW Tau (QSO); Markarian 421 UMa (BL LAC); 3C273 Vir (QSO); 3C279 Vir (QSO); Markarian 3C273 Vir; Markarian 509 Aqr; OJ+287 Cnc (BL LAC); W Com(1) (BL LAC); W Com(2) (BL LAC); BL Lac(1) (BL LAC); and BL Lac(2) (BL LAC).

Of these, the following have TA charts on the accompanying CD-ROM:

OJ+287 Cnc; W Com; BL Lac; and BW Tau.

Of those in the foregoing list the following also have BAAVSS charts on the accompanying CD-ROM:

BL Lac; BW Tau; Markarian 421 UMa; 3C273 Vir; and Markarian 509.

There are also TA charts of the additional AGNs 3C66A and 3C371 on the accompanying CD-ROM.

We have covered a lot of ground in this book; from gently pulsating stars in our own backyard to massive cosmic explosions way across the Universe. I hope that you have found this tour guide interesting and feel inclined to embark on your own programme of variable star observation. There is plenty to do and I know you will experience a great deal of enjoyment from this activity. You will also be joining in with your fellow amateur and professional astronomers and be performing real astronomical research.

Glossary

absolute magnitude The magnitude any celestial body, such as a star, would appear to have if it were placed at the standard distance of 10 parsecs (32.6 light years) from the Earth.

accretion disk A disk of material formed around an astronomical body, the material having been extracted from another astronomical body.

AGN The abbreviation usually used for 'active galactic nuclei', which are the nuclei of galaxies showing quasar-like activity due to the presence of a supermassive black hole at their centres.

altitude In astronomical usage, the angular distance separating a given astronomical body from the horizon.

amplitude Strictly in science and mathematics this term is used for half the total variation of any quantity between its maximum and its minimum values. However, in the field of variable stars – especially in the amateur arena – it is usually taken to mean the total variation in magnitude between maximum and minimum.

apparent field of view Relating to particular designs of eyepiece, it is the angle through which the observer's eye would have to swivel in order to see from one side of the visible field of view to the other.

apparent magnitude The brightness of a celestial body, expressed on the magnitude scale, as it appears in our sky. Its value is determined by the absolute magnitude and the distance it is from us. We can only measure apparent magnitudes and can only calculate values of absolute magnitude if we can also determine the distance to the celestial body in question (often a value of absolute magnitude can be inferred from other pieces of information, such as spectral type, etc., from which the star can be assigned to a recognised category).

aperture synthesis The process of combining the outputs of two or more telescopes to mimic a larger one. The main advantage of this method is in obtaining much finer resolution than would be obtained from any one of the telescopes.

astronomical unit (AU) The mean Earth–Sun distance; 1 AU = 150 million kilometres.

azimuth The companion co-ordinate to altitude; the angular distance of a celestial body after being projected vertically downwards onto the horizon plane, measured eastwards from the north cardinal point.

barycentre The centre of mass of any co-orbiting system; the point in space about which the two or more components of the system orbit.

Big Bang The generally accepted scenario for the creation of the Universe.

binary star Two stars mutually gravitationally bound into a co-orbiting system.

BL Lac objects Galaxies with active nuclei having very bland spectra, dominated by continuum emission resulting from synchrotron processes.

black body A theoretical concept: a perfect emitter and absorber of radiation, obeying well-defined physical laws.

black hole A region of space surrounding a collapsed star where the acceleration due to gravity exceeds the speed of light. In other words, the region inside the black hole is completely cut off from our Universe.

Bok globule Small, relatively dense cloud of interstellar material, from which new stars are born.

cataclysmic variable star Star showing explosive or nova-like outbursts due to matter being transferred from one star to another via the accretion disk surrounding the other (usually a white dwarf).

CCD This stands for 'charged coupled device'; the electronic detector in modern imaging systems.

CCD photometry Photometry carried out using a CCD camera as the brightness-measuring device.

celestial equator The projection of the Earth's equator onto the celestial sphere.

celestial poles The projection of the Earth's rotation poles onto the celestial sphere.

celestial sphere An imaginary huge sphere, having the Earth situated at its centre, on to which the co-ordinate systems used in astronomy are projected.

Cepheid variable stars Pulsating stars which are variable in brightness and which obey a specific law relating their brightnesses with their periods of variation. They vary their brightnesses with clockwork regularity and precision.

Chandrasekhar limit The maximum mass (1.4 solar masses) for a star to be able to exist as a white dwarf.

chromatic aberration A defect of an optical system where the images of a point source in different wavelengths of light (different colours) are brought to differing positions of focus.

chromosphere The layer of matter situated above the photosphere of the Sun (or other stars) from which spicules and prominences originate.

comparative photometry The procedure of using comparison stars to determine the brightness of one of unknown brightness (magnitude) by means of a photometer.

comparison stars Stars of constant and known brightnesses (apparent magnitudes) used for the purpose of determining the brightness (apparent magnitude) of a given (usually variable) star.

Glossary

cosmology The study of the birth, nature, and evolution of the entire Universe.

declination The angular distance separating a celestial body from the celestial equator, measured along a great circle that passes through the celestial body and the celestial poles.

Doppler effect The apparent change in wavelength of the radiation received from an emitting source caused by the relative motion between the receiver and the source.

dwarf nova A star which erupts semi-periodically, often once every few months, and brightens by a number of magnitudes. Its eruptions are much less extreme than those of novae but happen very much more frequently.

eclipse The passage of one celestial body behind another or into the shadow cast by it.

eclipsing variable star One of the classifications of variable star, the light variations being caused by one star eclipsing another as they co-orbit each other.

eruptive variable star One of the classifications of variable star, pertinent to those that undergo erratic, and often relatively minor, sudden changes in light output.

field of view In astronomical usage, the amount of sky one can see in one go through a telescope or binoculars fitted with a particular eyepiece. This is more exactly called 'the real field of view' of the eyepiece, as opposed to the 'apparent field of view' of the eyepiece.

globular star cluster A spheroidal grouping of mostly old stars. The stars in a typical cluster can number from a hundred thousand to about a million.

GRB The normal abbreviation for 'γ-ray bursts (gamma-ray bursts)'. Most examples, the bursts lasting 1 second or more, are thought to be produced by the swallowing of a star that is going supernova by the black hole created at its centre (a hypernova). Those bursts lasting under a second have as-yet undetermined origins.

Hertzsprung–Russell diagram A plot of the luminosity (or magnitude) against spectral type (or temperature) for a collection of stars.

Hubble's law An observed relationship between the recessional velocity of a galaxy and its distance.

hypernova A colossal outpouring of energy caused by the core of a dying supermassive star collapsing to form a black hole, which then quickly consumes the whole star.

interstellar medium The material lacing the spaces between the stars.

ionise The act of turning electrically neutral atoms into ions. Active form: 'ionisation'.

ions Atoms that have more or less electrons than the number of protons in their nuclei. Consequently they have a net electrical charge.

Julian Day Number Often used in astronomy, especially in variable star work, the number of days elapsed since noon on 4713 January 1 BC. This system avoids the complications that arise in the civil calendar system.

Kepler's laws Empirical laws describing how one celestial body orbits another. They have been given good mathematical foundations.

L_1 point The usually used shorthand for the 'inner Langrangian point', the Langrangian point that lies on a line between two co-orbiting massive bodies.

Langrangian point Any point in a complex gravitational field (the result of two or more massive bodies in proximity) at which a body will be in a state of equilibrium, balanced between the attractions of each of the massive bodies.
light-curve A graphical plot of a star's brightness variations with time.
light year A unit of measurement equal to the distance a pulse of light would travel in one year.
long-period variable star A variable star having certain characteristics, particularly cyclic (though often somewhat irregular) brightness variations of periods of months to years. This term is often used synonymously with Mira variable.
luminosity The brightness of a celestial body.
magnitude A logarithmic scale of the brightness of a celestial body. Apparent magnitude is a measure of the apparent brightness, while absolute magnitude refers to the body's real brightness.
Milky Way The apparent band of radiance that crosses the sky owing to the concentration of stars along the plane of our Galaxy.
Mira variable star A semi-regular red giant star.
nadir The point on the celestial sphere that lies vertically below the observer (and thus through the other side of the Earth). It is thus antipodal to the zenith.
nebula A vast cloud of gas and dust in space formed by a concentration of the interstellar medium.
neutrino An elementary particle liberated by nuclear reactions in the core of the Sun and of other stars, as well as during supernova explosions. This particle has a tiny mass and interacts only weakly with matter.
neutron star A star that has become crushed under its own weight to such an extent that its electrons have fused with its protons to make neutrons. Within the star the pressure may be high enough to transform neutrons into exotic forms of matter (quarks).
nova An eruptive outburst in which a star dramatically increases its brightness: the result of mass transfer processes between the stars of a close binary system.
nucleosynthesis The formation of new elements by nuclear reactions.
occultation The passage of one body behind another, as a result of their respective motions.
open star cluster A loose grouping of stars arising from their common birth place in the interstellar medium.
parallax The apparent shift in the position of a celestial body against the backdrop of more distant bodies caused by the motion of the observer.
parsec A unit of measurement equal to 3.26 light years.
period In variable star work, the time spanning two brightness maxima (or sometimes two brightness minima if that is more appropriate) of a variable star. It is also used to mean the time taken for any regular cyclic change, such as the orbit of one body around another.
photoelectric photometer A now nearly obsolete (at least in astronomy) device used for measuring brightnesses, superseded by the CCD.

Glossary

photoelectric photometry Measures of the brightnesses of sources using a photoelectric photometer.

photometer The general name given to any device who purpose is to measure the brightness of a particular source (stars, for instance).

photon A 'particle' of electromagnetic radiation.

photosphere The visible surface of the Sun or of another star.

PMT The usual abbreviation for 'photomultiplier tube', the active device used in that type of photometer.

precession A slow shift in the orientation of the spin-axis of a celestial body, or in the points (nodes) of maximum and minimum separation in the orbit of one body around another.

proper motion The apparent shift in the position of a star because of its real motion through space.

pulsar A rapidly pulsing radio source. These have been identified with neutron stars.

pulsating variable stars Stars whose brightness variations are caused by their outer layers expanding and contracting.

Purkinje effect A visual effect with a number of characteristics, in particular causing the brightnesses of red stars to be overestimated when comparing them with stars that are less red.

quasar A small and extremely powerful source of electromagnetic radiation. Quasars are now identified with the cores of active galaxies.

R Coronae Borealis star A specific type of variable star, which undergoes sudden drops in its brightness in connection with clouds of carbon particles forming in the space immediately surrounding it.

RR Lyrae star A specific type of variable star. It is actually a type of pulsating variable with certain well-defined characteristics, including a clockwork-like precise period of brightness variations.

radial velocity The component of a celestial body's velocity in the line of sight of the observer.

range In variable star work it is taken to mean the total extent of any brightness changes in an astrovariable, usually taken over the entire time it has been observed. 'Range' is often a much more appropriate (and accurate) term than 'amplitude'.

recurrent novae Novae which have been seen to erupt more than once during the time they have been observed.

red giant A bloated star with a cooler photosphere than that of our Sun. Its redness is a consequence of its lower photospheric temperature.

redshift The lengthening in the wavelengths of the spectral lines from a celestial body caused by its recessional radial velocity (*see also* Doppler effect).

resolution The discerning of fine details in an image.

right ascension The companion co-ordinate of declination. It is the projection of the celestial body's position onto the celestial equator, measured from the vernal equinox in an easterly direction along the celestial equator. It can be measured as an angle or in units of sidereal time.

Glossary

Roche limit The distance from a planet at which any large body (such as a natural satellite) would break up owing to the shearing forces caused by the planet's gravitational field.

Roche lobe For a pair of co-orbiting stars the regions around each star where the gravity of that star is dominant.

semiregular variable star A variable star which has some degree of periodicity in its light variations but is not as clockwork-regular as stars of the Cepheid and RR Lyrae types.

Seyfert galaxy A type of galaxy having a nucleus that is optically very bright. It shows certain characteristics that place it in a category between quasars and ordinary galaxies.

sidereal time A system of time measurement based upon the apparent motions of the stars (in fact, caused by the rotation of the Earth).

solar constant The total energy flux received per unit area per unit time at the radius of the Earth's orbit from the Sun.

solar wind A stream of electrified particles ejected from the Sun that spreads radially outwards through the Solar System.

spectral type The classification of a star based upon the characteristics of its spectrum.

standstill An interval when the normal light variations of a given variable star are interrupted and it remains at virtually constant brightness.

supergiant An extremely massive star. The bluest (highest photospheric temperature) examples are ten times the diameter of our Sun and the reddest (lowest photospheric temperature) examples can be a thousand times the Sun's diameter.

supernova A colossal stellar explosion, resulting either from the collapse of the core of an old massive star, or from runaway nuclear reactions caused by the transfer of matter from a star to a white dwarf to which it is bound in a close binary system.

symbiotic stars A category of variable stars whose light variations arise because of the transfer of matter from one star to another.

synchrotron radiation The emission of polarised electromagnetic radiation that arises from electrons spiralling along magnetic field lines.

transit The passage of one body in front of another as a result of their respective motions.

variable star A star whose brightness changes over time, aside from any changes that occur on timescales of millions of years or more arising from the evolution of the star.

white dwarf A stellar relic of superdense material (of the type known as electron degenerate) typically with a mass ranging from a little less to a little more than our Sun, packed into a sphere a few thousand kilometres in diameter.

X-ray bursters Sources producing intense bursts of X-radiation, usually due to matter being transferred from one star to a massive and compact object, or to the accretion disk around it.

zenith The point immediately overhead at a given observation site.

Resources

Here is a list of resources to help you get started, and subsequently pursue, your study and observation of astrovariables. It is necessarily limited but by making use of your local library (and especially making use of the interlibrary loan service), browsing publishers' catalogues and the advertisements for equipment suppliers, perusing astronomical magazines and periodicals and, above all, by surfing the Internet while making use of a search engine such as Google.com, you can uncover plenty more!

Books

Astronomy In Depth, Gerald North, Springer-Verlag, 2003.
Advanced Amateur Astronomy, Gerald North, Cambridge University Press, Second edition, 1997.
Astronomical Equipment for Amateurs, Martin Mobberley, Springer-Verlag, 1999.
The Art and Science of CCD Astronomy, edited by David Ratledge, Springer-Verlag, 1997.
A Practical Guide to CCD Astronomy, by Patrick Martinez *et al.*, Cambridge University Press, 1997.
The New CCD Astronomy: How to Capture the Stars with a CCD Camera in Your Own Backyard, by Ron Wodaski, New Astronomy Press, 2002.
Observing Variable Stars: A Guide for the Beginner, David Levy, Cambridge University Press, 1989.
Spectral Classification & Multicolour Photometry, edited by C. H. Fehrenbach, Kluwer Acandemic Publisher, 2002.
Light Curves of Variable Stars: A Pictorial Atlas, edited by C Sterken and C. Jaschek, Cambridge University Press, 1993.
Light Curve Modelling of Eclipsing Binary Stars, edited by E. F. Milone, Springer-Verlag, 1993.

Eclipsing Binary Stars: Modelling and Analysis, by Joseph Kallrath *et al.*, Springer-Verlag, 1999.

Mass-Losing Pulsating Stars and Their Circumstellar Matter: Observations and Theory (Astrophysics and Space Sciences Library vol. 283), edited by Y. Nakado *et al.*, Kluwar Academic Publishing, 2003.

Cataclysmic Variable Stars: How and Why They Vary, Coel Hellier, Springer-Praxis, 2001.

Cataclysmic Variable Stars (Cambridge Astrophysics), edited by Brian Warner, Cambridge University Press, 2003.

Historical Supernovae and Their Remnants (International Series on Astronomy and Physics, vol. 5), by F. Richard Stephenson and David A. Green, Oxford University Press, 2002.

The Extravagant Universe: Exploding Stars, Dark Energy, and the Accelerating Cosmos, by Robert Kirshner, Princeton University Press, 2002.

Cosmic Catastrophes: Supernovae, Gamma-Ray Bursts, and Adventures in Hyperspace, by Craig Wheeler, Cambridge University Press, 2000.

The Biggest Bangs: The Mystery of Gamma-Ray Bursts, the Most Violent Explosions in the Universe, by Jonathan I. Katz, Oxford University Press, 2002.

Supernovae and Gamma-Ray Bursts, edited by Mario Livio *et al.*, Cambridge University Press, 2001.

Active Galactic Nuclei, by Julian Henry Krolik, Princeton University Press, 1998.

Some national associations/societies which have sections devoted to variable star observing, or are entirely devoted to this activity

In the UK the premier national amateur astronomical society is the *British Astronomical Association (BAA)*. The address is:

Burlington House, Piccadilly, London W1J 0DU, England.
Telephone: 020–7734 4145
Fax: 020–7439 4629
Home page: http://www.britastro.org/

Of particular note to us is the Variable Star Section of the BAA (known as the BAAVSS). Its web page contains a vast resource of additional information, such as details of ongoing research programmes, a list of all objects for which the BAAVSS has a database, the most up-to-date charts, etc.:

http://www.britastro.org/vss

There is a separate group operating from the UK but with an international membership which is set up with separate observing sections like the BAA, which has strong links with the BAA, and which has particularly strong professional–amateur links and is internationally respected. This is: *The Astronomer Magazine (TA)*. TA exists mainly as a clearing house for observations and it is a very

National associations and societies

valuable resource for up-to-date information on observations and discoveries in variable star and other aspects of observational astronomy. It has no one main address but the Secretary and the leaders of each of the observing sections are best contacted via the addresses, telephone numbers and/or email addresses given on the 'contact details' link on the TA home page:

http://theastronomer.org/index.html

The United States of America has an observing society which specialises in variable stars: the *American Association of Variable Star Observers (AAVSO)*. The address is:

25, Birch Street, Cambridge, MA 02138, USA.
Home page: http://www.aavso.org/

In Japan there is *The Variable Star Network (VSNET)*, of Kyoto University. The address is:

Kitashirakawa-Oiwake-cho, Saykyo-ku, Kyoto 606–8502, Japan.
Home page: http://www.kusastro.kyoto-u.ac.pp/vsnet

An association in Germany devoted solely to variable stars is the *Bundesdeutsche Arbeitsgemeinshaft für Veränderliche Sterne (BAV)*.
The address is:

Munsterdamm 90, 12169 Berlin, Germany.
Home page: http://thola.de/bav.html

The equivalent association in France is the *Association Française des Observateurs d'Etoiles Variables (AFOEV)*.
The address is:

Observatoire Astronomique de Strasbourg, 11 rue de l'Université, 67000 Strasbourg.
Home page: http://cdsweb.u-stasbg.fr/afoev/

Observers of the southern skies are served by the *Astronomical Society of South Australia (ASSA)*, whose Secretary can be contacted at:

GPO Box 199, Adelaide, SA 5001, Australia.
Home page: http://www.assa.org.au/info/

and by the *Royal Astronomical Society of New Zealand (RASNZ)*, whose Secretary can be contacted at:

PO Box 3181, Wellington, New Zealand.
Home Page: http://www.rasnz.org.nz/

Some useful web-addresses

Some more websites offering resources
The General Catalogue of Variable Stars Research Group http://www.sai.msu.sv/groups/cluster/gcvs/gcvs/
The Catalogue and Atlas of Cataclysmic Variables, Living Edition http://icarus.stsci.edu/downes/cvcat/
Information Bulletins on Variable Stars http://www.konkoly.hu/IBVS.html

Image processing programs
Mira   http://www.axres.com/
AIPWIN   http://www.willbell.com/aip4win/AIP.htm
MaxIm DL   http://www.cyanogen.com
IRIS   http://astrosurf.com/buil/us/iris/iris.htm
Starlink   http://star-www.rl.ac.uk/
IRAF   http://iraf.noao.edu/iraf-homepage.html

CCD cameras
Commercial cameras
Starlight XPress   http://www.starlight-xpress.co.uk/
SBIG   http://www.sbig@sbig.com
Apogee   http://www.apogee-ccd.com/

Do-it-yourself cameras
Audine   http://astrosurf.com/audine/
Cookbook   http://www.wvi.com/~rberry/cookbook.htm

Spectroscopy
Buil   http://astrosurf.com/buil/us/spe1/spectro1.htm
Spectr'aude   http://astrosurf.com/buil/us/spectro8/spaude_us.htm
Maurice Gavin   http://www.astroman.fsnet.co.uk/spectro.htm

Astronomical alert services
IAU circulars   http://cfa-www.harvard.edu/iau/services/ Subscriptions.html
TA circulars   http://www.theastronomer.org/subscription.html
Sky and Telescope   http://skyandtelescope.com/observing/proamcollab /astroalert

Index

AB Aurigae 106, 107
absolute magnitude 3–4, 211
absorption spectra 90, 94, 148
accretion disk 162, 162–163, 164,
    176–179, 180–181, 183, 186, 187, 193,
    200, 208, 209, 211
active galactic nuclei 1, 203–210, 211
active galaxies see active galactic nuclei
AC Herculis 145
ACV stars see rotating variable stars
ACVO stars see rotating variable stars
ACYG stars 123, 144
adaptive optics unit 69
AGN see active galactic nuclei
AIP4WIN software 75
Algol 155–157, 159, 162
Algol-type stars see EA stars
Alnilam 56
Alnitak 56
α Canis Majoris see Procyon
α Cygni stars see ACYG stars
α Ursae Minoris see Polaris
American Association of Variable Star
    Observers (AAVSO) 12–13, 14, 219
AM Herculis 186
AM Herculis stars see AM Her stars
AM Her stars 185–186
Andromeda, Great Galaxy in see M31
anomalous Cepheids see BL Boo stars
antiblooming drain 61, 62, 63, 70
antinode, definition of 124
apparent visual magnitude 2, 3, 4, 211

AR subtype stars 158
Arbour, Ron 197
Arcturus 97, 163
Argelander, Friedrich 7, 155
Armstrong, Mark xi, 197, 203
autocollimating eyepiece 51

Baade, Walter 130
back-illuminated CCDs 61, 64
Bailey, S. I. 131
Balmer lines 94, 95
band spectra 92
barycentre 154, 153–154, 173, 212
Bayer, Johann 133
BCEP stars 10, 101, 144
BCEPS stars 123, 144
Belopolsky, A. A. 121–122
BeppoSAX satellite 201, 202
β Canis Majoris stars 10
β Cephei stars see BCEP stars
β Geminorum see Pollux
β Lyrae 159
β Lyrae stars see EB stars
β Persei see Algol
Betelgeuse xii, 1, 62, 96, 137–139
Bevis, John 192
bias frame 71, 72, 73
binary star systems 152, 153–171,
    172–189, 193, 212
binned mode 65
binoculars 20, 22, 29, 36–38, 39, 57
Blazhko effect 132

Index

BL Boo stars 10, 123
BL Boötis stars *see* BL Boo stars
BL Lacertae objects (BL Lac objects) 1, 206, 207, 209, 212
black body 85–86, 112, 212
black holes 174, 188, 199–200, 202, 207, 208–209, 212
blinking (search technique) 170, 198
blueshift, spectral 95, 96–98
Bok globules 74, 102–103, 212
Boles, Tom xi, 197, 198–199, 203
bright giant stars 96, 111
bright spot (on accretion disk) 162, 163, 180, 181, 182
British Astronomical Association (BAA) i, xi, 12, 15, 14–16, 18, 24, 76, 77, 108, 133, 134, 138, 144, 145, 146, 147, 164, 166, 177, 178, 183, 186, 194, 197, 204, 207, 218
brown dwarf stars 94, 105
Buczynski, Denis xi
Buil, Christian 75, 76
BYDRA stars *see* rotating variable stars

3C 48 203, 204
3C 273 203–204
3C 279 204
calibration frames 70, 71–73
carbon cycle 83, 84, 112, 131
Carlin, Nils Olof 51
Cassegrain telescopes 29, 33, 34, 35–36, 47–48, 49, 50
cataclysmic variable stars 1, 8, 95, 172–189, 190–210, 212
catadioptric telescopes 20, 22–23, 33, 49, 50, 63, 64, 69
CCD photometry i, 39, 59–80, 212
CEP stars 101, 123, 128–131, 133, 134, 212, 216
CEP(B) stars 123, 128–131
Cepheid instability strip 101, 118, 119, 127, 128
Cepheid variable stars *see* CEP stars, CEP(B) stars, DCEP stars, CW stars, *and* CWB stars
Chandra satellite 201, 202
Chandrasekhar limit 190, 212
charge coupled devices (CCDs) i
charts *see* finder charts

Cheshire eyepiece 51
chromosphere 88, 90, 94, 148, 212
classical Cepheid variable stars *see* Cepheid variable stars
classical novae *see* N stars *and* NA stars *and* NB stars *and* novae
CNO cycle *see* carbon cycle
collimation of telescopes 42–51
colour–magnitude diagram *see* H–R diagram
coma 22
common envelope phase 173, 187
compact radio sources 205
comparison stars 29, 52, 53, 54–55, 56, 58, 70, 137, 141, 150, 212
comparison star sequence method 54–55
Compton Gamma Ray Observatory 200–201
contact binary star systems 162
continuous spectra 88
constellations, list of 5–6
convective zone 82, 88
corona 82, 88
Cortini, Giancarlo 197
Crab Nebula 192–193
CW stars 123, 128–131
CWA stars 123, 128–131
CWB stars 123, 128–131
Cygnus X-1 199–200

D subtype stars 158
dark current 61
dark frame 71, 72–73, 74
DCEP stars 101, 123
DCEPS stars 123, 128–131
degeneracy pressure *see* electron degeneracy pressure
degenerate matter *see* electron degenerate matter
δ Cephei 121–123, 128, 130
δ Scorpii 149
δ Scuti stars *see* DSCT stars
δ 163
detached binary star systems 162
detector quantum efficiency (DQE) 60
diaphragm *see* stopping-down telescopes
digital cameras 62
dispersion 94, 95

Index

distance modulus 3–4
DN stars *see* dwarf novae
Doppler effect 95, 155, 213
DQ Herculis stars *see* DQ Her stars
DQ Her stars 185, 186
DS subtype stars 158
DSCT stars 101, 123, 144, 145
DSCTC stars 123
Dumbbell Nebula 113
Durlevich, Dr O.V. xi, 9
DW subtype stars 158
dwarf novae 9, 101, 172, 178, 181, 175–185, 186, 213
dwarf stars 97, 144, 145, 156, 176

E stars *see* eclipsing binary systems
EA stars 158, 161
Earth ix, 3, 4, 102, 117, 157
EB stars 158
eclipsing binary systems 1, 8, 146–149, 156, 181, 180–183, 186, 213
eclipsing dwarf novae 180–183
eclipsing variable stars *see* eclipsing binary systems
Eddington, Sir Arthur 126, 127, 128–129
effective temperature 112
ELL stars *see* rotating variable stars
electron degeneracy pressure 116, 164, 190
electron degenerate matter 116, 117, 131, 164, 167, 173
emission spectrum, definition of 90
EP stars 158
ε Canis Majoris 96
eruptive variable stars 1, 141–143, 146–151, 213
estimating star brightnesses i, 39–58
η Carinae 131, 150
η Carinae Nebula 103, 104
Evans, Reverend Robert 197
event horizon 199
exit pupil 25–28, 29, 37
extended radio sources 205
extrinsic variable stars 8
eyepieces 21, 29–31, 33
EW stars 158, 161

Fabricus, David 133
field of view 20–21, 22, 29–31, 37, 57, 211, 213
filters *see* photometric filters

finder charts i, 13–16, 52, 53, 67, 106, 143, 146, 148–149, 158–159, 168, 170, 183–185, 187–189, 197, 199, 210, 229
finderscopes *see* finder telescopes
finder telescopes 29, 67
FKCOM stars *see* rotating variable stars
flare stars 9, 150
flat field 71–72, 73, 74
flickering 175
flip-mirror system 67–68
Foulkes, Steven 197
fractional method 55
frame transfer CCDs 61, 63–64
front-illuminated CCDs 61, 65
FU Orionis stars *see* FU stars
FU stars 9, 106
full frame CCDs 61
fundamental mode, definition of 124

GAIA probe 4
Galaxy ix, 99, 102, 116, 128, 130, 153, 164, 192, 195, 203, 208, 209, 214
γ-ray bursters *see* γ-ray bursts
γ-ray bursts 1, 200–203, 213, 229
γ Cassiopeia 142, 149
γ Cassiopeia stars *see* GCAS stars
γ Doradus stars *see* GDOR stars
γ Orionis 97
GCAS stars 101, 149–150
GDOR stars 144
General Catalogue of Variable Stars (GCVS) 10–11, 106, 121, 123, 133, 137, 157, 185
General Catalogue of Variable Stars Research Group xi, 9, 10, 11, 135, 141, 149, 158, 169, 188, 220
giant stars 1, 97, 98, 101, 110, 111, 117, 118, 137, 143, 149, 187, 200 *see also* red giant stars
globular star clusters 129, 130, 131, 132, 213
Goodricke, John 121, 131, 155
gravitational equipotentials *see* gravitational equipotential surfaces
gravitational equipotential surfaces 160–163
gravitational potential 159, 161
gravitational potential energy 159, 160
GRB *see* γ-ray bursts
Great Nebula in Orion *see* Orion Nebula

Index

Greek alphabet 7
GS stars 158

H–R diagram 98–99, 100, 101, 105,
    110–111, 112, 116, 117, 118, 118, 119,
    121, 123, 126–127, 128, 129, 196,
    213
Harvard Designations (HD) 7–8
helium flash 117, 118, 131
Herschel, John 137
Hertzsprung, Ejnar 98
Hertzsprung–Russell diagram *see* H–R
    diagram
Hevelius, Johann 133
Hewitt, Dr Nick xi
Hipparcos satellite 4, 130
Historiola Mirae 133
Holwarda, Phocylides 133
Hubble, Edwin 129, 130
Hubble Heritage Team xi, 104, 109,
    206
Hubble Space Telescope 104, 109, 115,
    208
Hurst, Guy xi
hydrostatic equilibrium 84, 87, 106, 121,
    125
hyperfine structure 95
hypergiant stars 141–143
hypernovae *see* γ-ray bursts

I stars 150
IA stars 150
IB stars 150
IC 418 115
image scale 20, 61, 63–65
inner Langrangian point *see*
    Langrangian point
IN stars *see* nebula variable stars
INA stars *see* nebula variable stars
INB stars *see* nebula variable stars
INS stars 151
INT stars *see* nebula variable stars
intense X-Ray sources *see* X-ray
    sources
interline transfer CCDs 61
intermediate polars 185–186
interstellar medium 100–103, 213
intrinsic variable stars 8
IN(YY) stars *see* nebula variable stars
IP Pegasi 183, 230
IRIS software 75, 76

irregular stars *see* I stars, IA stars, *and* IB
    stars
irregular variable stars 9
IS stars 150
ISA stars 151
ISB stars 151
IX Velorum 169

James, Nick x, 14, 16–19, 62, 74, 76, 77,
    181, 182, 183, 203
Joule–Kelvin effect 126
Julian Day Numbers 14, 16–19, 213

K subtype stars 158
KE subtype stars 158
Kepler, Johann 154
Kepler's Laws of Planetary
    Motion 154–155, 213
KW subtype stars 158

L_1 point *see* Langrangian point
L stars 123, 139, 143
Lambda Boötis stars *see* LBOO stars
Langrangian point 161–163, 164, 173,
    213, 214
Large Magellanic Cloud 195
laser collimators 51
Laurie, Stephen 197
Laws of Planetary Motion *see* Kepler's
    Laws of Planetary Motion
LB stars 123, 143
LBOO stars 145
LBV stars 123, 143
LC stars 101, 123, 143, 144
Leavitt, Henrietta 128
Levy, David 137
light-curves i, 14, 16–19, 54, 106, 108,
    123, 132, 134, 135–137, 138, 139–140,
    143, 144, 145, 146, 147, 148, 149, 156,
    158, 159, 164, 166, 167, 168, 169–170,
    176, 177, 178, 180, 181, 182, 183–185,
    186, 187, 192, 194, 197, 204, 207, 210,
    214, 229
line profile 95
line spectra *see* spectral lines
long-period variable stars 9, 133, 214
longitudinal oscillation, definition
    of 124
low-mass stars 113
luminosity classes, stellar (explanation
    of) 95, 96–98

Index

M stars 101, 123, 133–137, 214
M1 *see* Crab Nebula
M13 130
M27 *see* Dumbbell Nebula
M31 129, 130
M42 *see* Orion Nebula
M57 *see* Ring Nebula
M81 194
M87 205, 206
magnitudes 2–3, 24–29, 37, 79, 211, 214
magnitude scale 2–3, 214
main sequence stars 97, 98, 101, 108, 110, 111, 112, 113, 116, 119, 121, 131, 167, 172, 173, 187
Maksutov telescopes 23, 33, 49, 50
mass–luminosity relation 110–111
Matthews, Thomas 203
Mazza, P. 197
Messier, Charles 192
Milky Way *see* Galaxy
Mintaka 56
Mira 133–135
Mira stars *see* M stars
Mobberley, Martin xi, 194, 203, 217
Montanari, Geminiano 155
mountings, telescope 23, 68–69, 198

N stars 163–168 *see also* novae
N81 108
NA stars 163–168 *see also* novae
NASA *see* National Aeronautics and Space Administration
National Aeronautics and Space Administration xi, 24, 104, 109, 115, 206
NB stars 163–168 *see also* novae
NC stars 163–168 *see also* novae
nebula variable stars 101, 106–108, 146, 150
neutrinos 119
neutron stars 120, 174, 188, 190–191, 193, 199, 200, 214, 215
Newton, Isaac 154
Newtonian telescopes 21–22, 23, 27, 29, 33, 34, 34–35, 44, 43–47, 49, 51, 69, 74
NGC 3132 115
NGC 3372 104
NGC 4527 197
NGC 6751 115
NL stars 169–170, 185

nodes, definition of 124
nomenclature 4–8
Norton's Star Atlas 52
Nova Cygni 1975 *see* V1500 Cygni
novae 1, 9, 18, 101, 153–171, 172, 193, 214 *see also* dwarf novae *and* recurrent novae
nova hunting 170–171
nova-like variable stars *see* NL stars
Nova Vulpeculae 1979 165, 166
NR stars 9, 167–168, 215

o Ceti *see* Mira
Open star clusters 103, 105, 214
Orion Nebula 102, 103
Orion variable stars *see* nebular variable stars
OS Andromedae 15
overtones, definition of 125

Paczynski, Dr Bohdan 201
parallax *see* trigonometrical parallax
parsec 3
period–luminosity law 128–129
period–luminosity relation *see* period–luminosity law
Pesci, Stefano 197
photoelectric effect 90
photometric filters 78–79
photometry 59–80, 183, 203, 212, 214, 215
photosphere 82, 90, 94, 97, 98, 106, 110, 112, 121, 122, 126–127, 128–129, 135, 148, 151, 161, 162, 187, 191, 215
Pickard, Roger xi
Pigott, Edward 131
planetary nebulae 113, 115, 116, 117, 119, 131, 135, 146–148, 174
Platt, Terry xi, 66
Pleiades 103, 105
PN stars 158
Pogson's step method 56
Polaris 128
polars 185–186
Pole star *see* Polaris
Pollux 137
population I stars, definition of 120
population II stars, definition of 120
population III stars, definition of 120
PPI cycle *see* proton–proton cycle
PPII cycle *see* proton–proton cycle

PPIII cycle *see* proton–proton cycle
pressure broadening 92, 95
Procyon 137
propeller effect 174
proton–proton cycle 83, 84, 87, 112, 121
PR Persei 144
PSR stars *see* rotating variable stars
pulsars *see* neutron stars
pulsating variable stars 1, 8, 121–132, 133–149, 215
Purkinje effect 56–57, 215
PVTEL stars 123, 145

quantum efficiency *see* detector quantum efficiency
quarks 190–191
quasars 1, 203–210, 215
quasi-stellar radio sources *see* quasars

R Boötis 18
R Coronae Borealis 146, 147, 148
R Coronae Borealis stars *see* RCB stars
R Cygni 7, 8
radiative zone 88
radio galaxies *see* active galactic nuclei
random walk 67–68
rapid irregular variable stars *see* IS stars, INS stars, ISA stars, *and* ISB stars
rapid pulsating hot subdwarf variable stars *see* RPHS stars
RCB stars 101, 146–149, 215
recurrent novae *see* NR stars
red giant stars xii, 1, 117, 118, 119, 143, 164, 173, 174, 180, 215 *see also* giant stars
redshift, spectral 95, 215
reflecting telescopes 20, 21–22, 23, 24, 26, 27, 28–29, 33, 34–36, 43–48, 49, 50, 51, 69, 74, 91, 95, 113, 129, 130, 163, 192, 194, 196, 201 *see also* Newtonian telescopes *and* Cassegrain telescopes
reflection (phenomenon in a binary star system) 157–158
reflectors *see* reflecting telescopes
refracting telescopes 20–21, 22–23, 27, 29, 48–49, 50, 165
refractors *see* refracting telescopes
ρ Cassiopeia 141–143
Rigel 96

Ring Nebula 113, 116–117
Roche lobes 158, 161–163, 173, 187, 216
Rosse, Lord (William Parsons) 192
rotating variable stars 8, 101, 151–152
Royal Greenwich Observatory (RGO) i, xi, 91, 151, 163, 165
RPHS stars 145
RR Cygni 7
RR stars 101, 123, 131–132, 133, 134, 215
RRAB stars 123, 131–132
RR(B) stars 123, 131–132
RRC stars 123, 131–132
RR Lyrae stars *see* RR stars, RR(B) stars, RRAB stars, *and* RRC stars
RR Tauri 106
RS Cygni 7, 8
RS stars 158
RT Cygni 7
RU Cephei 138
RU Cygni 7
Russell, Henry Norris 98
RV stars 123
RVA stars 123, 145
RVB stars 123, 145
RZ Cygni 7

Samus, Dr Nikolai N. xi, 9
Sandage, Allan 203, 204
Santa Barbara Instruments Corporation (SBIG) 66
Schaefer, Bradley E. 24, 25, 27, 28–29
Schmidt–Cassegrain telescopes 22, 23, 33, 34, 49, 50, 63, 64, 65, 69, 198
Schmidt, Maarten 204, 206
Schmidt–Newtonian telescopes 23, 33, 49
Schwarzschild radius 199
S Cygni 7
SD subtype stars 158
SDOR stars 101
seeing disk 64
Seyfert, Carl 205
Seyfert galaxies 1, 205–206, 209, 216
semi-detached binary star systems 162, 176
semi-regular variable stars *see* SR stars, SRA stars, SRB stars, SRC stars, SRD stars, *and* SRS stars
sequence charts 13–16
Shapley, Harlow 129, 130, 131

Index

singularity 199
Sirius 2
Sky Atlas 2000.0 52
Small Megallanic Cloud 108, 109, 128
SN 1991 T 197
SN 1993 J Ursae Majoris 182
SN 1994 I 196
solar constant 85
Space Telescope Science Institute (STScI) xi
spectral lines, explanation of 92–93, 94
spectral response (of detectors) 2, 60, 61, 78–79
spectral types (of stars), explanation of 92–93, 94
SR stars xii, 9, 137–141, 216
SRA stars 123, 137–141
SRB stars 123, 137–141
SRC stars 101, 123, 137–141
SRD stars 123, 137–141
SRS stars 137–141
SS Cygni 7, 175, 176, 177, 178
SS Cygni stars *see* dwarf novae *and* UGSS stars
ST Cygni 7
standard candles, definition of 129
star brightnesses *see* magnitudes
star clusters 98, 103, 130 *see also* globular star clusters *and* open star clusters
star diagonal 50
star-hopping 52–53, 67
Starlight Xpress Ltd. xi, 65, 66
Starry Night Pro software 53
Stebbins, Joel 155
Stefan–Boltzmann law 86–87, 96, 112
Sternberg Astronomical Institute xi, 9
stopping-down telescopes 28–29, 53
subdwarf stars 97, 146
subgiant stars 97, 144, 155–157, 164, 176
Sun ix, 3, 4, 82, 81–88, 90, 95, 97, 100, 110, 111, 112, 117–118, 127, 135, 150, 151, 154
sunspots 151
supergiant stars 74, 96, 97, 101, 111, 118, 119, 137, 143, 144, 145–146, 195–196, 216
supermassive black holes *see* black holes
supermaxima *see* superoutbursts

supernovae 1, 120, 191–199, 202, 216
supernova hunting 197–199
supernova remnants 192–193
superoutbursts 176
SU Ursae Majoris 177
SU Ursae Majoris stars *see* UGSU stars
SXARI stars *see* rotating variable stars
SXPHE stars 123, 146
SX Phoenicis stars *see* SXPHE stars
symbiotic stars 1, 9, 167, 172–189, 216
synchronous rotation 157
synchrotron emission *see* synchrotron radiation
synchrotron radiation 192, 216
SZ Cygni 7

T Coronae Borealis 168
T Cygni 7
T Pyxis 167
T Tauri 108
T Tauri variable stars 106, 108 *see also* nebular variable stars
telecompressor lenses 22, 65
telescope mounts *see* mounting telescopes
The Astronomer Magazine (TA) xi, 12, 14, 15, 16–19, 171, 197, 199, 203, 218–219
thermal equilibrium 84–87, 102
track-and-accumulate exposures 69–70
transverse oscillations, definition of 124
trigonometrical parallax 3–4, 129, 214
triple-α process 131
TT Cygni 7
TU Cygni 7
type I supernovae 193, 195
type Ia supernovae 193
type Ib supernovae 193
type II supernovae 195–197
type A eruptions (of a dwarf nova) 183
type B eruptions (of a dwarf nova) 183

U Cygni 7
U Geminorum 18, 175–176
U Geminorum stars *see* dwarf novae
U Monocerotis 145
UG stars *see* dwarf novae
UGSS stars 176, 177

227

Index

UGSU stars 176, 177
UGZ stars 176, 177, 178
universal law of gravitation 154
Uranometria 2000.0 52
UU Herculis stars 139 *see also* SRS stars
UV Ceti stars see UV stars
UV stars 150
UX UMa stars 185
UX Ursae Majoris stars *see* UX UMa stars

V Boötis 138
V Ursae Minoris 138
V335 Orionis 7
V336 Orionis 7
V351 Orionis 106
V586 Orionis 106
V705 Cassiopeia 18, 166, 167
V770 Cassiopeia 144
V1057 Cygni 106, 107
V1419 Aquilae 167
V1500 Cygni 163, 164, 166
variable star classification, explanation of 8–10
Vega 2, 4, 95, 163
very slow novae *see* NC stars
vignetting 22, 31–36, 57, 69
Villi, Mirko 197
Virgo A 205 *see also* M87
visual magnitude *see* apparent visual magnitude
visual response of eye 2
Vogel, H. C. 155
VY Scl stars 185
VY Tauri 108

W Ursae Majoris stars *see* EW stars
W Virginis stars *see* Cepheid variable stars
WC stars 150 *see also* WR stars
WD stars 158
webcams 62

white dwarf stars 98, 101, 116–117, 118, 119, 131, 145, 146, 163, 164, 167, 168, 173, 174, 175, 176, 179, 180, 181, 181, 182, 183, 185–186, 187, 188, 190, 193, 199, 216
WN stars 150 *see also* WR stars
Wolf–Rayet stars *see* WR stars *and* WC stars *and* WN stars
Worraker, W. J. xi, 76, 77, 181, 182, 183
WR stars 93, 150, 158

X stars 188
X-ray bursters 1, 216
X-ray sources 8, 188–189, 192, 199–200
XB stars 188
XF stars 188
XI stars 188
XJ stars 188
XM stars 188
XND stars 188
XNG stars 188
XP stars 188
XPR stars 188
XPRM stars 188

Y Lyncis 138

Z Andromedae 186
Z Andromedae stars *see* ZAND stars
Z Camelopardalis 176, 177, 178
Z Camelopardalis stars *see* UGZ stars
Z Cygni 7
ZAMS *see* Zero Age Main Sequence
ZAND stars 186–188
Zeeman effect 95
Zero Age Main Sequence 108, 118
ZZ stars 101, 123, 146
ZZA stars 123, 146
ZZB stars 123, 146
ZZ Ceti stars *see* ZZ stars *and* ZZA stars *and* ZZB stars *and* ZZO stars
ZZO stars 123, 146

The accompanying CD-ROM

(constructed by Nick James)

The accompanying CD-ROM is compatible with a wide range of computers and all the data on it can be accessed using your favourite web browser. The pages are designed to use a very basic form of HTML which should be supported by practically every browser and operating system. The disk itself is mastered using the ISO 9660 level 2 file system since this is universally supported by those operating systems which support CD-ROM drives.

Start this CD-ROM as you would any other on your computer (for instance, by clicking on the CD-ROM drive icon that appears in 'My Computer'). You will see a number of folder icons displayed. They contain all the data files and you can ignore them for all normal purposes. The file icon labelled 'index.html' is the one you want. Click on it to open the contents page and you will see 'links' arranged under headings:

Charts
TA chart catalogue
TA galaxy catalogue
BAAVSS charts

Lightcurves
Lightcurves from BAAVSS observations

Miscellaneous
Various images
Variability types
The first UK Gamma Ray Burst detection

The accompanying CD-ROM

Movies
Simulation of a CV
Timelapse movie of an IP Peg eclipse

Just click on these 'links' to open the pages and you can continue to navigate just the same as you would if you were navigating through websites on the Internet.